THE
COSMIC CALCULATOR

(A Vedic Mathematics Course for Schools)

India's Scientific Heritage

General Editor: Dr L M Singhvi

3

Editorial Panel

THE
COSMIC CALCULATOR

(A Vedic Mathematics Course for Schools)

Book 1

KENNETH WILLIAMS
MARK GASKELL

Foreword by
L.M. SINGHVI
Formerly High Commissioner for India in the UK

MOTILAL BANARSIDASS PUBLISHERS
PRIVATE LIMITED • DELHI

First Indian Edition: Delhi, 2002

ISBN: 81-208-1862-8
ISBN: 81-208-1871-7 (Set)

Also available at:
MOTILAL BANARSIDASS
41 U.A. Bungalow Road, Jawahar Nagar, Delhi 110 007
8 Mahalaxmi Chamber, 22 Bhulabhai Desai Road, Mumbai 400 026
120 Royapettah High Road, Mylapore, Chennai 600 004
236, 9th Main III Block, Jayanagar, Bangalore 560 011
Sanas Plaza, 1302 Baji Rao Road, Pune 411 002
8 Camac Street, Kolkata 700 017
Ashok Rajpath, Patna 800 004
Chowk, Varanasi 221 001

Illustrations by David Williams
Cartoons by John Fuller

Printed in India
BY JAINENDRA PRAKASH JAIN AT SHRI JAINENDRA PRESS,
A-45 NARAINA, PHASE-I, NEW DELHI 110 028
AND PUBLISHED BY NARENDRA PRAKASH JAIN FOR
MOTILAL BANARSIDASS PUBLISHERS PRIVATE LIMITED,
BUNGALOW ROAD, DELHI 110 007

FOREWORD

Through trackless centuries of Indian history, Mathematics has always occupied the pride of place in India's scientific heritage. The poet aptly proclaims the primacy of the Science of Mathematics in a vivid metaphor:

यथा शिखा मयूराणां, नागानां मणयो यथा।
तद्वद् वेदांगशास्त्राणाम् गणितं मूर्ध्नि स्थितम्॥

—वेदांग ज्योतिष*

Mathematics is universally regarded as the science of all sciences and "the priestess of definiteness and clarity". J.F. Herbert acknowledges that "everything that the greatest minds of all times have accomplished towards the comprehension of forms by means of concepts is gathered into one great science, Mathematics". In India's intellectual history and no less in the intellectual history of other civilisations, Mathematics stands forth as that which unites and mediates between Man and Nature, inner and outer world, thought and perception.

Indian Mathematics belongs not only to an hoary antiquity but is a living discipline with a potential for manifold modern applications. It takes its inspiration from the pioneering, though unfinished work of the late Bharati Krishna Tirthaji, a former Sankaracharya of Puri of revered memory who reconstructed a unique system on the basis of ancient Indian tradition of mathematics. British teachers have prepared textbooks of Vedic Mathematics for British Schools. Vedic mathematics is thus a bridge across centuries, civilisations, linguistic barriers and national frontiers.

Vedic mathematics is not only a sophisticated pedagogic and research tool but also an introduction to an ancient civilisation. It takes us back to many millennia of India's mathematical heritage. Rooted in the ancient Vedic sources which heralded the dawn of human history and illumined by their erudite exegesis, India's intellectual, scientific and aesthetic vitality blossomed and triumphed not only in philosophy, physics, astronomy, ecology and performing arts but also in geometry, algebra and arithmetic. Indian mathematicians gave the world the numerals now in universal use. The crowning glory of Indian mathematics was the invention of zero and the introduction of decimal notation without which mathematics as a scientific discipline could not have made much headway. It is noteworthy that the ancient Greeks and Romans did not have the decimal notation and, therefore, did not make much progress in the numerical sciences. The Arabs first learnt the decimal notation from Indians and introduced it into Europe. The renowned Arabic scholar, Alberuni or Abu Raihan, who was born in 973 A.D. and travelled to India, testified that the Indian attainments in mathematics were unrivalled and unsurpassed. In keeping with that ingrained tradition of mathematics in India, S. Ramanujan, "the man who knew infinity", the genius who was one of the greatest

---

* Lagadha, Verse 35

mathematicians of our time and the mystic for whom "a mathematical equation had a meaning because it expressed a thought of God", blazed new mathematical trails in Cambridge University in the second decade of the twentieth century even though he did not himself possess a university degree.

I do not wish to claim for Vedic Mathematics as we know it today the status of a discipline which has perfect answers to every problem. I do however question those who mindlessly deride the very idea and nomenclature of Vedic mathematics and regard it as an anathema. They are obviously affiliated to ideological prejudice and there ignorance is matched only by their arrogance. Their mindset were bequeathed to them by Macaulay who knew next to nothing of India's scientific and cultural heritage. They suffer from an incurable lack of self-esteem coupled with an irrational and obscurantist unwillingness to celebrate the glory of Indian achievements in the disciplines of mathematics, astronomy, architecture, town planning, physics, philosophy, metaphysics, metallurgy, botany and medicine. They are as conceited and dogmatic in rejecting Vedic Mathematics as those who naively attribute every single invention and discovery in human history to our ancestors of antiquity. Let us reinstate reasons as well as intuition and let us give a fair chance to the valuable insights of the past. Let us use that precious knowledge as a building block. To the detractors of Vedic Mathematics I would like to make a plea for sanity, objectivity and balance. They do not have to abuse or disown the past in order to praise the present.

Dr. L.M. Singhvi
Formerly High Commissioner for India in the UK

> "*Every move of every ant and elephant, the starsand galaxies,*
> *at every moment are in perfect precision in space and time,*
> *managed by Vedic Mathematics—the mathematics of pure knowledge.*"
> Maharishi Mahesh Yogi

INTRODUCTION

Mathematics has three main branches: Arithmetic, Algebra and Geometry.

Arithmetic deals with numbers, of which there are many types. Numbers have various properties and can be combined in various ways.

Algebra deals with symbols, often the letters of the alphabet. These also have their own special properties and can be combined.

Geometry deals with shapes and forms: points, lines, surfaces and solids. There are many different kinds of form and they have many interesting properties.

These three branches develop in a structured way and cover all areas of Mathematics.

Some topics in mathematics combine two or three of these branches.

At the beginning of this book you will see a list of Sutras (or formulae) and a list of Sub-Sutras (sub-formulae) which will be a great help in your mathematics work. The word "Sutra" means "thread", and we will find that these formulae link up many areas of mathematics. In fact they describe the way your mind works: they tell you how to go about answering questions. You will be introduced to these Sutras one by one and you will find they are *written in italics* in the book.

You will find that the methods of Vedic Mathematics are very easy and straightforward. So simple, in fact, that you may want to do most of the calculations in your head, mentally. You are encouraged to do this because the Vedic system is a system of mental mathematics. But you should also enjoy the mathematics so you should not struggle to do everything in your head. Except for the mental tests at the beginning of the lesson you are free to use pencil and paper if you like. Have fun!

✸ Please also note the following:

> Important facts or laws are contained in a rectangular white box like this one, this will help you to refer back to them and to revise your work.

There is an **answer** book which has answers to the exercises.

Where you see a star like the one at the top of this page, or a number at the beginning of a line this indicates that there is something for you to do. But you will only generally find answers to the numbered instructions in the answer book.

Important words are written in **bold** when they are introduced for the first time.

VEDIC MATHEMATICS SUTRAS

1	एकाधिकेन पूर्वेन Ekādhikena Pūrveṇa	*By One More than the One Before*
2	निखिलं नवतश्चरमं दशतः Nikhilam Navatascaramam Dasatah	*All from 9 and the Last from 10*
3	ऊर्ध्वतिर्यग्भ्यां Ūrdhva Tiryagbhyām	*Vertically and Crosswise*
4	परावर्त्य योजयेत् Parāvartya Yojayet	*Transpose and Apply*
5	शून्यं साम्यसमुच्चये Sūnyam Sāmyasamuccaye	*If the Samuccaya is the Same it is Zero*
6	आनुरूप्ये शून्यं अन्यत् (Ānurūpye) Sūnyamanyat	*If One is in Ratio the Other is Zero*
7	संकलन व्यवकलनाभ्यां Saṅkalana Vyavakalanābhyām	*By Addition and by Subtraction*
8	पूरणापूरणाभ्यां Pūraṇāpūraṇābhyām	*By the Completion or Non-Completion*
9	चलनकलनाभ्याम् Calana Kalanābhyām	*Differential Calculus*
10	यावदूनं Yāvadūnam	*By the Deficiency*
11	व्यष्टिसमष्टिः Vyaṣṭisamaṣṭih	*Specific and General*
12	शेषाणयङ्केन चरमेण Sesāntyadvayamantyam	*The Remainders by the Last Digit*
13	सोपान्त्यद्वयमन्त्यं Sopāntyadvayamantyam	*The Ultimate and Twice the Penultimate*
14	एकन्यूनेन पूर्वेन Ekanyunena Purvena	*By One Less than the One Before*
15	गुणितसमुच्चयः Guṇakasamuccayah	*The Product of the Sum*
16	गुणकसमुच्चयः Guṇakasamuccayah	*All the Multipliers*

SUB-SUTRAS

1	त्रानुरूप्येण Ānurupyeṇa	*Proportionately*
2	शिष्यते शेषसंज्ञः Sisyate Seṣasaṃjñah	*The Remainder Remains Constant*
3	त्राधमाधेनान्त्यमन्त्येन Ādyamādyenānantyamantyena	*The First by the First and the Last by the Last*
4	केवलैः सप्तकं गुरायात् Kevalaih Saptakam Guṇyāt	*For 7 the Multiplicand is 143*
5	वेष्टनम् Veṣṭanam	*By Osculation*
6	यावदूनं तावदुनं Yāvadūnam Tāvadūnam	*Lessen by the Deficiency*
7	यावदूनं तावदूनीकृत्य वर्गं च योजयेत् Yāvadūnam Tāvadūnikrtya Vargañca Yojayet	*Whatever the Deficiency lessen by that amount and set up the Square of the Deficiency*
8	त्रन्त्ययोर्दशकेऽपि Antyayordasake'p	*Last Totalling 10*
9	त्रन्त्ययोरेव Antyayoreva	*Only the Last Terms*
10	समुच्चयगुणितः Samuccayagunnitah	*The Sum of the Products*
11	लोपनस्थापनाभ्यां Lopanaṣṭāpanābhyām	*By Alternate Elimination and Retention*
12	विलोकनं Vilokanam	*By Mere Observation*
13	गुणितसमच्चुयः समुच्चयगुणितः Guṇitasamuccaya Samuccayagunitah	*The Product of the Sum is the Sum of the Products*
14	ध्वजाङ्ड Dhvajāṅka	*On the Flag*

CONTENTS

1

ARITHMETIC

The first of the three branches of mathematics is arithmetic, which deals with number.

The number 1 represents unity and is the first number. Unity and wholeness are everywhere.

The system of Vedic Mathematics is based on 16 formulas (or Sutras). For example *By One More Than the One Before* (*Ekadhikena Purvena* in the Sanskrit language).

✳ Find this Sutra in the list at the front of this book.

Starting from number 1 all whole numbers can be generated by using *By One More Than the One Before*: 2 is one more than 1; 3 is one more than 2 and so on. We will see many other applications of this simple formula as we go on.

Whole numbers are also called **natural numbers**.

> Every number is different from every other number.
> Like people, every number has its own special and unique properties.

✳ Write the numbers 1 to 10 vertically down the left-hand side of your page leaving about 2 lines of space between each.

✳ For each number discuss with a partner where you have seen this number illustrated.
For example, for number 7 you may think of the 7 days of the week or the 7 colours of the rainbow. Write your observations down on your list.

✳ Discuss some properties of different numbers.
For example, how can you distinguish between odd and even numbers,
    what is special about the number one,
    what is special about the number ten?
Think of as many possibilities as you can.

EXERCISE 1

Which number is one more than:

a 6	**b** 17	**c** 9	**d** 19	**e** 21	**f** 30
g 99	**h** 199	**i** 300	**j** 401	**k** 9,999	**l** 309
m 6,789	**n** 7,829,009	**o** 9,699	**p** 9,998	**q** 78,979	

There is also a sutra called *Ekanunena Purvena.*

What do you think this means? (Look at your list of sutras)

EXERCISE 2

Go back to Exercise 1, but this time write down the number that is one *less* than the number given.

"Even the highest and farthest reaches of modern Western mathematics have not yet brought the Western world even to the threshold of Ancient Indian Vedic Mathematics."

Professor A. De Morgan (1806-71)

MATHEMATICIAN AND LOGICIAN

2

DIGIT SUMS AND THE NINE POINT CIRCLE

The word **digit** means the single figure numbers: the numbers 1 to **9 and** 0.
Sum means addition.

> The **digit sum** of a number is found by adding the figures in the number.

 EXAMPLE 1

Find the digit sum of   **a** 17   **b** 401.
a For 17 we get $1 + 7 = \underline{8}$.   **b** For 401 we get $4 + 0 + 1 = \underline{5}$.

 EXERCISE 1

Find the digit sums of the following numbers:

a 16   **b** 27   **c** 32   **d** 203   **e** 423   **f** 30103

g the digit sum of a 2-figure number is 8 and the figures are the same. What is the number?

h the digit sum of a 2-figure number is 9 and the first figure is twice the second. What is it?

EXAMPLE 2

Find the digit sum for 761.

For 761 we get $7 + 6 + 1 = 14$. And as 14 is a 2-figure number we add the figures in 14 to get $1 + 4 = 5$.

So the digit sum of 761 is 5.

The digit sum is found by adding the digits in a number, and adding again if necessary.

This means that any natural number of any size can be reduced to a single digit: just add all the digits, and if you get a 2-figure number, add again.

 EXERCISE 2

Find the digit sums of:

a 37	**b** 461	**c** 796	**d** 6437	**e** 3541
f 5621	**g** 4978272	**h** 673982741		

THE NUMBER NINE

The number nine is very important in the Vedic system. As we will see it has many remarkable properties which make it very interesting and very useful.

THE NINE POINT CIRCLE

The nine point circle is a circle whose edge is divided into 9 equal parts.

✷ First draw a circle, with your compass set on a radius of about 4 cm. Leave plenty of space all around your circle.

As accurately as you can, divide your circle into thirds as shown. Ask your teacher for a hint on how to do this.

✴ Now divide each third into three and number the circle as shown.

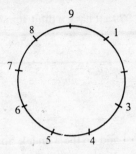

This is known as the 9 point circle. We are now going to continue numbering round the circle.

✴ Where shall we put the number 10?

It seems reasonable to put 10 after 9, where we have already got number 1.

We can continue numbering around the circle as shown below.

✴ Where will the number 19 go?

Continue numbering around the circle up to 30.

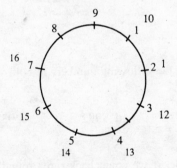

1 Write down all the numbers you get on the 'branch' that begins with 1, e.g. 1,10,19 etc.
2 What do you notice about their digit sums?
3 Predict the next 3 numbers on this branch.
4 What are their digit sums?
5 What can you say about the digit sums of all the numbers on this branch?
6 Do the same thing for two other branches. What do you notice?
7 Starting at 3 what do you have to add on to get back to the 3 branch?
8 What would you have to add on to any one number to get back to the branch you started at?

> Adding 9 to a number does not affect its digit sum:
>
> so 7, 70, 79, 97, 979 all have a digit sum of 7 for example.

 EXAMPLE 3

Find the digit sum of 39409.

We can **cast out the nines** and just add up the 3 and 4. So the digit sum is 7.

Or using the longer method we add all the digits: 3+9+4+0+9 = 25 = 7 again.

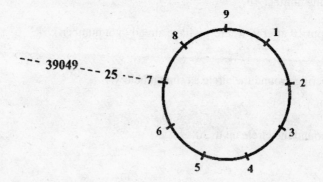

EXERCISE 3

Try working out the digit sums of the following numbers in both ways to see if you get the same answer. Show your working clearly.

a 39   **b** 93   **c** 993   **d** 9993   **e** 9329   **f** 941992

Looking again at the 9 point circle, if we count backwards round the circle we see that since 0 comes before 1 it is logical to put zero at the same place as 9. In terms of digit sums 9 and 0 are equivalent and this also explains why we can cast out nines.

> As digit sums, 9 and 0 are equivalent.

This **casting out the nines** can be used in another way:

> Any group of figures in a number that add up to 9 can also be "cast out".

EXAMPLE 3

Find the digit sum of   **a** 24701   **b** 21035

a In 24701 we see 2 and 7 which add up to 9. We can therefore cast them out and add up
only the other figures. 4+0+1 = 5.

b In 21035 we can see that 1, 3 and 5 add up to 9 so we cast these out.
The sum of the remaining figures is 2 so this is the answer.

 EXERCISE 4

Calculate the digit sums of the following:

a 39   **b** 94   **c** 95   **d** 995   **e** 9995   **f** 634

g 724   **h** 534   **i** 627   **j** 9653   **k** 36247   **l** 9821736

YOUR LUCKY NUMBER

✳ Write down your date of birth and find its digit sum.

✳ You can also find the digit sum of your phone number and your house number.

 DIGIT SUM GAME   For 2, 3 or 4 people.

You will need 36 cards made from thin card or paper.
Four of them should be numbered 1, four numbered 2 and so on up to 9.

Place the cards, spread out with the numbers downwards, on the table.

The player who starts picks a card. If it is 9 the card is kept, otherwise another card is picked.
If the total of the two cards is 9 they are kept, if the total is more than 9 the cards are replaced
on the table and the next player has a go, if the total is less than 9 another card is picked until
the total is either 9 or over 9.

So each player picks cards until they get 9 (in which case they keep the cards),
or until their total is over 9 (in which case they replace the cards on the table).

Everyone should be allowed to see the cards that are turned over.
You should try to remember the cards that are put back so that you can get your total of 9 when
it is your turn.

The player with the most nines at the end is the winner.

DIGIT SUM PROBLEMS

 EXERCISE 5

Find a 2-figure number with a digit sum of:

a 6 where the figures are equal
b 6 where the 1st figure is double the 2nd
c 7 where the difference between the figures is 3   (two answers)
d 7 where one figure is a 4   (two answers)
e 6 where both figures are odd   (three answers)
f 5 where the figures are consecutive #   (two answers)
g 9 where the figures are consecutive   (two answers)
h 9 where one figure is double the other   (two answers)
i 8 where the number is below 20
j 1 where the first figure is a 4

 # Consecutive means one after the other. e.g. 6 and 7 are consecutive (or 7 and 6).

 EXERCISE 6

Find a 2-figure number with a:

a digit sum of 5 that ends in 2
b digit sum of 8 that is between 50 and 70   (two answers)
c digit sum of 7 where the figures are equal
d digit sum of 2 and one of the figures is a 3   (two answers)
e digit sum of 1 that ends in 3
f digit sum of 3 that ends in 8
g digit sum of 1 where the figures are equal
h digit sum of 2 that ends in 7
i digit sum of 6 where one figure is a 2   (two answers)
j digit sum of 3 which is greater than 80   (two answers)
k digit sum of 2 where one figure is a 4   (two answers)
l digit sum of 5 where both figures are odd   (three answers)
m digit sum of 2 where the number is even   (five answers)
n digit sum of 8 where the number is odd   (five answers)
o digit sum of 3 where the figures differ by 6   (two answers)

> *"It is India that gave us the ingenious method of expressing all numbers by means of ten symbols, each symbol receiving a value of position as well as an absolute value; a profound and important idea which appears so simple to us now that we ignore its true merit. But its very simplicity and the great ease which it has lent to all computations put our arithmetic in the first rank of useful inventions..."*
>
> Marquis de Pierre Simon Laplace (1749-1827)
>
> FRENCH MATHEMATICIAN

3

LARGE NUMBERS

In a large number like **1234** the value of each figure depends on its position:

thousands	hundreds	tens	units
1	2	3	4

The 4, being at the extreme right is 4 units,
the 3 is in the tens position, and so represents 3 tens, or 30,
the 2 is 2 hundreds
and the 1 is 1 thousand.

If a particular column is empty we put 0.
For example, 3030 is recognised as three thousand and thirty.

Our number system uses the number **10** as a **base**.
We will call the numbers 10, 100, 1000 etc. **base numbers**.

RESTRUCTURING NUMBERS

We need to be flexible about the way numbers are shown in this system.
Sometimes it is necessary to restructure a number.

EXAMPLE 1

In the number **33** we have 3 tens and 3 units

tens	units		tens	units
3	3	=	2	13

Here **33** is rewritten as **2** tens and **13** units.

tens	units		tens	units
3	3	=	1	23

and here it becomes **1** ten and **23** units.

 EXAMPLE 2

It is also easy to convert a number like

tens	units
4	16

into the simpler form by, in this case, adding the 1 to the 4 to get **56**.

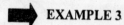 **EXAMPLE 3**

Similarly

hundreds	tens	units		hundreds	tens	units
4	3	2	=	4	2	12

Also

hundreds	tens	units		hundreds	tens	units
4	3	2	=	3	13	2

 EXERCISE 1

Put into their simplest form (we write h t u for hundreds, tens and units):

a h t u
  1 4 13

b h t u
   5 23

c h t u
   5 1 22

d h t u
      9 15

e h t   u
  3 14 8

f h t   u
  1 25 7

g h t   u
  3 13 13

h h t   u
     9 9 10

Copy the following and insert the correct number where there is a question mark:

i   t u   t u   t u
  44 = 3 ? = 2 ? = 1 ?

j   h t u   h t u   h t u   h t u
  555 = 5 4 ? = 5 ? 25   = 4 ? 5 = 4 ? 25

k   h t u   h t u   h t u   h t u
  741 = 6 ? 1 = 5 ? 1 = 7 2 ? = 5 ? 21

READING AND WRITING LARGE NUMBERS

A **thousand** is ten hundreds. It is written 1,000 or 1000.

If you arranged ten centimetre cubes in a row and then put ten of these rows together to make a square, you would have 100 cubes.

If now you had ten identical layers of 100 cubes on top of each other you would have ten hundreds or 1000 cubes.

 EXAMPLE 4

The number **54,321** says **fifty four thousand, three hundred and twenty one**.

"**54**" is pronounced first, then "**thousand**" (where the comma is), then "**321**".

 EXERCISE 2

Write the following in words, try to get the spelling right:

a 76,462   **b** 18,703   **c** 6,480   **d** 90,008   **e** 111,234   **f** 41,070

g 3,800   **h** 309,511   **i** 88,055   **j** 123,456   **k** 30,000   **l** 120,300

Sometimes you may see large numbers written out without the commas. But since the comma is always three figures from the right-hand end it is easy to see where it should be.

And sometimes a space is left instead of a comma.
So you might see 76 000 or 76,000 or just 76000. Any of these forms might be used in this book.

Now can we do the opposite? Given a number in words, can we write it in figures?

 EXAMPLE 5

Seventeen thousand nine hundred and forty six would be written as 17,946.
Note that the word "thousand" in this example shows you where the comma goes.

 EXAMPLE 6

Seven hundred and eighty three thousand and sixty six.

First write down seven hundred and eighty three, 783, then the comma, then 66.
This gives us 783,066.
Note carefully the extra 0 which must be inserted. Without this 0 the number would be 78366 which says seventy eight thousand three hundred and sixty six: which is a different number.

There must always be 3 figures after the thousands and this may involve adding 0 or 00 or 000.

 EXERCISE 3

Write the following in figures:

a Seven thousand eight hundred and twenty.

b Thirty three thousand four hundred and twelve.

c Fifty one thousand eight hundred and ninety.

d Thirteen thousand six hundred and seventy seven.

e Thirty four thousand eight hundred and four.

f Three hundred and forty three thousand seven hundred and eleven.

g Four hundred and ten thousand two hundred and eighty.

h Eight hundred and eighteen thousand and seventy four.

i Fifty seven thousand and ninety nine.

j Eighty thousand nine hundred and nine.

k Three hundred thousand seven hundred and forty six.

l Two hundred and twenty thousand and one.

m Ninety thousand.

n Four hundred and four thousand three hundred and three.

MILLIONS

This system can be extended to include millions, billions and so on.

A million is a thousand thousand. It is written in figures as 1,000,000 or 1000,000.
This is a huge number .

✸ Can you think of a way of showing 1 million, like we had for 1 thousand?

Perhaps you could look at a sheet of graph paper. Find out how many of the smallest squares there are on the whole sheet. Look for an easy way of doing this.

✸ How many sheets of graph paper would you need to have 1 million small squares?

EXAMPLE 7

Write 23,456,789 in words.

Seeing 2 commas we know we have millions here (23 millions).
The middle part is the thousands so we have 456 thousands.

So we can write
twenty three **million** four hundred and fifty six **thousand** seven hundred and eighty nine.

The Sutra we have been using here is *Lopanasthapanabhyam* which means *By Alternate Elimination and Retention*. This is because we are taking each of the three parts of the number in turn: when we look at the millions we ignore the rest of the number, then we ignore the first and last parts when we look at the thousands, and so on.

EXAMPLE 8

Write 103,040,005.

This says one hundred and three **million** forty **thousand** and five.

EXAMPLE 9

Write 17,000,800.

Seventeen **million** eight hundred   (there are no thousands here).

EXERCISE 4

Write in words:

a 6,202,304   **b** 10,005,070   **c** 85, 085,200   **d** 111,900,007

e 108,108,108   **f** 99,000,333   **g** 2,020,200   **h** 999,001,070

EXERCISE 5

Write in figures:

a Eight million eight thousand three hundred and four.

b Ninety million eight hundred and seven thousand and one.

c Three hundred million four hundred thousand and seventy seven.

d Twelve million five hundred and sixty five thousand six hundred and fifty six.

e Four million and forty.

f Four hundred and thirty two million.

BILLIONS

A **billion** is a thousand million. That is 1,000,000,000.

 EXAMPLE 10

Write in words   70,030,200,000.

This would be <u>seventy billion, thirty million, two hundred thousand.</u>

 EXAMPLE 11

Write the number   three billion, eight hundred and eight million and one.

This is <u>3,808,000,001</u>.

 MAKE A NUMBER   A game for 2 to 5 players

Put 3 of the white number cards face up in a row in the middle of the table.
Put the remaining number cards and the red pack face down on the table.

Players take turns to take a card from each pack and attempt to make the number requested with the 4 white cards at their disposal (they need not use all 4 cards).

If successful, the player keeps one number card of their choice and puts the red card at the bottom of its pack, otherwise both cards they picked go to the bottom of their packs.

Players should discuss whether an answer is correct or not and ask the teacher if necessary.

The game ends when the white pack has gone.

The player with the most cards at the end is the winner.

"The Sutras (aphorisms) apply to and cover each and every part of each and every chapter of each and every branch of mathematics including arithmetic, algebra, geometry . . . In fact, there is no part of mathematics, pure or applied, which is beyond their jurisdiction."

Sri Bhāratī Kṛṣṇa Tīrthajī (1884-1960)

DISCOVERER OF VEDIC MATHEMATICS

4

DIGIT SUM CHECK

Now we know how to calculate digit sums we can look at some useful applications. One very common use of digit sums, in which we can decide if a number is divisible by another number, will be examined in a later chapter. In this chapter we will see a method of checking the answers to addition or subtraction sums. We will also see a new method of subtraction.

ADDITION FROM RIGHT TO LEFT

Before we do so, we need to revise our method for writing down simple additions from right to left. In a later chapter we will learn how to do calculations both written and mentally from left to right.

 EXAMPLE 1

Find 3451 + 432.

```
  3 4 5 1        We line the numbers up with the units under the units.
+   4 3 2        There are no 'carries' so we simply add in each column:
  3 8 8 3        1+2=3;  5+3=8;  4+4=8;  3+0=3.
```

EXERCISE 1

Solve the following addition sums, setting them out in a similar way to the example.

a 43	**b** 21	**c** 123	**d** 401	**e** 4321	**f** 2102
23 +	34	201 +	121	511 +	127
—	42 +	—	77 +	—	330 +
	—		—		—

g 33+21   **h** 111+212   **i** 203+21   **j** 1414+123+ 212   **k** 234120+54302+43

SUMS INVOLVING "CARRIES"

 EXAMPLE 2

5 6
1 6 +
7 2
  ̄

In this example when we add 6 and 6 we get 12 so we write down 2 in the units column and 'carry' 1. We then add this carried 1 to 5+1 and write 7 in the tens column.

 EXAMPLE 3

97
88
185
  ̄

Here, similarly when we add 7 and 8 we get 15, so we write down 5 and carry 1. This 1 is then added to 9+8 and we write down 18.

 EXERCISE 2

Add the following. Leave plenty of room on either side of the sum for later calculations.

a 43	**b** 37	**c** 56	**d** 131	**e** 867	**f** 83
28 +	28 +	29 +	271 +	67 +	77 +
—	—	—	—	—	—

g 38+48   **h** 74+69   **i** 777 +136   **j** 2563+1723

THE DIGIT SUM CHECK

Having solved some simple addition problems, we can now learn how to perform a simple check using digit sums to see if they are correct.

This check comes under the Vedic Sutra *Gunitasamuccayah Samuccayagunitah*, which means *The Product of the Sum is the Sum of the Products*. The **product** of some numbers normally means the result of multiplying them together.

In its present application the formula reads *The Digit Sum of the Sum is the Sum of the Digit Sums*.

 EXAMPLE 4

Find 32 + 12 and check the answer using digit sums.

$$
\begin{array}{cc}
32 & 5 \\
\underline{12} + & \underline{3} + \\
44 & 8
\end{array}
$$

Here, the digit sum of 32 is 5 (3 + 2 = 5) and the digit sum of 12 is 3. The sum (the total) of the digit sums is 5 + 3 = 8. If we have done the sum correctly, the digit sum of the answer should also be 8. 4 + 4 = 8; so according to this check the answer is probably correct.

 EXAMPLE 5

Add 365 and 208 and check the answer.

$$
\begin{array}{cc}
365 & 5 \\
\underline{208} + & \underline{1} + \\
573 & 6 \\
1
\end{array}
$$

We get 573 for the answer.
We find the digit sums of 365 and 208 .
These are 5 and 1, and their total is 6.
Finally as the digit sum of 573 is also 6 the answer is confirmed.

 EXAMPLE 6

Add 77 and 124 and check.

$$
\begin{array}{cc}
77 & 5 \\
\underline{124} + & \underline{7} + \\
201 & 3 \\
\overline{11}
\end{array}
$$

This is similar except that when we add 5 and 7 we put down not 12 but 3, the digit sum of 12.

---

1 Now go back to exercise 2 and check your answers using the 'digit sum check'. If you find any mistakes, go back and do your corrections! Lay the sums out as shown above.

 EXERCISE 3

Add the following and check your answers using the digit sums:

a	35	b	56	c	35	d	52	e	456	f	188
	47 +		27 +		59 +		24 +		333 +		277 +

g 78+87   **h** 66+48   **i** 555+77   **j** 823+37   **k** 3760+481

EXERCISE 4

Which of the sums below are wrong? Use the digit sum check to find out.

a 43	**b** 91	**c** 58	**d** 55	**e** 46	**f** 88	**g** 438
75 +	19 +	76 +	39 +	47 +	77 +	238 +
118	110	124	94	94	165	676

h 516	**i** 444	**j** 46	**k** 911	**l** 79	**m** 53	**n** 83
615 +	878 +	71	119 +	89 +	676 +	38 +
1131	1312	38 +	1030	168	729	131
		154				

 EXAMPLE 7

Check the following sum: 279   The check is: 9
                             121 +   4 +
                             490   4 which confirms the answer.

However if you check the addition of the original sum you will find that it is incorrect!
This shows that the digit sum method does not always find an error. It usually works but not always.
We will be meeting other checking devices later on.

SUBTRACTION

You probably already know how to subtract numbers but in this section you will be shown a very easy method of subtracting numbers that you have probably not seen before.

 EXAMPLE 8

Find 35567 – 11828.

We set the sum out as usual: 35567
Then starting on the **left** we subtract in each column.   11828 –
3 – 1 = **2**, but before we put **2** down we check that in   2
the next column the top number is larger.
In this case 5 is larger than 1 so we put **2** down.

In the next column we have 5 – 1 = 4, but looking in the third
column we see the top number is not larger than the bottom   3 5¹5 6 7
(5 is less than 8) so instead of putting 4 down we put **3** and the   11828 –
other 1 is placed *On the Flag*, as shown so that the 5 becomes 15.   **2 3**   ☞

So now we have $15 - 8 = 7$. Checking in the next column
we can put this down because 6 is greater than 2.
In the fourth column we have $6 - 2 = 4$, but looking at
the next column (7 is smaller than 8) we put down only
3 and put the other one *On the Flag* with the 7 as shown.

$$3\ 5^15\ 6^17$$
$$\underline{1\ 1\ 8\ 2\ 8\ -}$$
$$\mathbf{2\ 3\ 7\ 3}$$

Finally $17 - 8 = \mathbf{9}$:

$$3\ 5^15\ 6^17$$
$$\underline{1\ 1\ 8\ 2\ 8\ -}$$
$$\mathbf{2\ 3\ 7\ 3\ 9}$$

 EXAMPLE 9

Find $535 - 138$.

Here we have $5 - 1 = 4$ in the first column,
but in the next column the figures are the same.
In such a case we must look one further column along.

$$5\ 3\ 5$$
$$\underline{1\ 3\ 8\ -}$$
$$\underline{}$$

And since in the third column the top number is smaller
we reduce the 4 to **3**:

$$5^13\ 5$$
$$\underline{1\ 3\ 8\ -}$$
$$\mathbf{3}$$

Then we proceed as before:

$$5^13^15$$
$$\underline{1\ 3\ 8\ -}$$
$$\mathbf{3\ 9\ 7}$$

We subtract in each column starting on the left, but before we put an answer down we look in
the next column.
If the top is greater than the bottom we put the figure down.
If not, we reduce the figure by 1, put that down and give the other 1 to the smaller number at
the top of the next column.
If the figures are the same we look at the next column to decide whether to reduce or not.

 EXERCISE 5

Subtract the following from left to right. Leave some space to the right of each sum so that you can
check them later.

a	444	b	63	c	813	d	695	e	765	f	504
	183 -		28 -		345 -		368 -		369 -		275 -

g	51	h	3456	i	7117	j	8008	k	5161	l	9876
	38 -		281 -		1771 -		3839 -		1838 -		6789 -

m 6 3 6 3   **n** 5 1 0 1 5   **o** 1 4 2 8 5   **p** 8 8 1 8 8   **q** 9 6 3 0 3 6 9
  3 3 8 8 –   2 7 9 8 6 –   7 1 4 8 –   1 9 1 8 9 –   3 6 9 0 9 6 3 –
  ———   ———   ———   ———   ————

CHECKING SUBTRACTION SUMS

 EXAMPLE 10

Find 69 – 23 and check the answer.

6 9	6	The answer is 46.
2 3 –	5 –	The digit sums of 69, 23 and 46 are 6, 5 and 1.
4 6	1	And 6 – 5 = 1, which confirms the answer because
		here we subtract the digit sums.

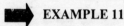 **EXAMPLE 11**

5 6	2
2 9 –	2 –
2 7	9

In this example, the digit sum of both 56 and 29 is 2 and the digit sum of 27 is 9.
We get 2–2 = 0, but we have already seen that 9 and 0 are equivalent as digit sums, so the answer is confirmed.

Alternatively, we can add 9 to the upper 2 before subtracting the other 2 from it: 11 – 2 = 9.
This is because adding 9 to any number takes you round the circle and back to where you started: 2 and 11 are equivalent in terms of digit sums.

 EXAMPLE 12

679	4
233 –	8 –
446	5

Here we have 4 – 8 in the digit sum check so we simply add 9 to the upper figure (the 4) and continue: 13 – 8 = 5, which is also the digit sum of 446.

? **EXERCISE 6**

a Check your answers to Exercise 5 by using the digit sum check.

b Which of the following sums are wrong? Use the digit sum check to find out.

i 48	**ii** 47	**iii** 78	**iv** 147	**v** 238	**vi** 876
23 –	25 –	69 –	69 –	47 –	389 –
25	22	12	78	181	487

5

NUMBER NINE

In our number system the number nine is the largest digit.

The number nine also has many other remarkable properties which make it extremely useful.

We have already seen that it can be used in finding digit sums, and that the digit sum of a number is unchanged if 9 is added to it or subtracted from it.

Now look at the 9-times table:

$$9 \times 1 = 9$$
$$9 \times 2 = 18$$
$$9 \times 3 = 27$$
$$9 \times 4 = 36$$
$$9 \times 5 = 45$$
$$9 \times 6 = 54$$
$$9 \times 7 = 63$$
$$9 \times 8 = 72$$
$$9 \times 9 = 81$$
$$9 \times 10 = 90$$
$$9 \times 11 = 99$$
$$9 \times 12 = 108$$

If you look at the answers you will see that in every case the digit sum is 9.

You may also see that if you read the answers as two columns the left column goes up, *By One More than the One Before* (1, 2, 3, . . .) and the right column goes down, *By One Less than the One Before* (9, 8, 7, . . .).

Next look at subtraction: choose a 2-figure number in which the first figure is greater than the second. Reverse the figures and subtract from the original number.

For example, choose 83, in which the first figure is greater than the second.

Reversing the figures gives 38 which we subtract from 83:

$$\begin{array}{r} 83 \\ 38 \; - \\ \hline \end{array}$$

1 Finish this sum and then find the digit sum of the answer.

✳ Choose a 2-figure number of your own and do a similar subtraction.

2 What is the digit sum of your answer?

3 Do the subtraction sum:

$$\begin{array}{r} 8\;3\;1 \\ 3\;1\;8 \; - \\ \hline \end{array}$$

4 What is the digit sum of the answer?

5 Find

$$\begin{array}{r} 7\;5\;6\;3 \\ 3\;7\;6\;5 \; - \\ \hline \end{array}$$ and the digit sum of the answer.

In the above two sums the figures are not reversed, but rearranged.

✳ Choose a number of any length you like, rearrange the figures (but make sure that the first figure is smaller than the first figure of the original number), subtract and find the digit sum of the answer.
You should find that it is 9 whatever number you choose.

6 Can you explain why this happens?

BY ADDITION AND BY SUBTRACTION

This Sutra is very useful when adding or subtracting numbers which end in 9 or 9's.

Suppose we want to add 9 to another number, say 44. We could count 44,45,46,47......etc.
9 times, being careful not to make a mistake.

✳ Can you think of another way of solving this problem?

If we think of 9 as being one less than 10 we could solve this problem using the sutra *By Addition and by Subtraction*. In other words we could add on 10 and then subtract 1.
Adding 10 to 44 gives 54, and taking 1 off gives 53.

Perhaps you would have done it this way or you may have just 'known' the answer, rather like knowing your tables. But we can use this idea for more complicated problems.

 EXAMPLE 1

Find   **a** 55 + 29   **b** 222 + 59   **c** 345 + 199

Each of these can be done in an easy way using the above formula.
a We could add 30 to 55 and take 1 off. So we can write the answer straight down:
     55 + 29 = 84.

b Here we add 60 and take 1 off: 222 + 59 = 281.

c 199 is 1 below 200 so we can add 200 to 345 and take 1 off: 345 + 199 = 544.

 EXAMPLE 2

Find   **a** 55 − 29   **b** 222 − 59   **c** 345 − 199

a Since 29 is 1 below 30 we can subtract 30 and add 1 back on: 55 − 29 = 26.

b Subtract 60 and put 1 back: 222 − 59 = 163.

c Subtract 200 and add 1: 345 − 199 = 146.

 EXAMPLE 3

Find   **a** 66 + 28   **b** 444 − 297

The same method can be applied here.
a We can add 30 to 66 and take 2 away: 66 + 28 = 94.

b We can subtract 300 and add 3 back on: 444 − 297 = 147.

 EXERCISE 1

> try these mentally

Do the following sums mentally:

Add 9 to: **a** 35   **b** 78   **c** 333   **d** 1009   **e** 1237   **f** 3007

g Subtract 9 from the above.

h 44 + 19   **i** 44 − 19   **j** 83 + 19   **k** 83 − 19   **l** 33 + 29   **m** 77 − 29

n 45 + 39   **o** 88 − 49   **p** 75 + 59   **q** 333 + 99   **r** 444 − 99   **s** 267 − 199

t 487 + 299   **u** 44 + 28   **v** 83 − 38   **w** 555 + 398   **x** 465 − 297   **y** 434 + 394

✳ With a partner practice some problems of this kind.
Take turns asking 5 questions, including adding/subtracting 9, 19, 29, 99, 199, 8, 38, 198, 97.

Don't use paper and try and get as many right as you can!

7 The 9-Squares Puzzle

You have the 9 numbers 1, 2, 3, 4, 5, 6, 7, 8, 9 and one number must go in each box above in such a way that each row and each column and both diagonals add up to the same number.

8 The 9-Dots Puzzle

Can you join the 9 dots together using only four straight lines, without taking your pencil off the paper?

9 The 9 - Nines Puzzle

You have 9 nines: **9, 9, 9, 9, 9, 9, 9, 9, 9** and you have as many + and − symbols as you like. What is the largest total you can make with these, using at least one of these symbols?
For example, can you get a bigger number than the answer to

$$999 + 9999 + 99?$$

And what is the smallest total you can get with these same symbols?

> "*L*ike the crest of the peacock,
> like the gem on the head of a snake,
> so is mathematics at the head of all knowledge."

from the Jyotish Vedanga

6

NUMBERS WITH SHAPES

Every number is a different expression of unity: it has its own unique qualities. All numbers fall into different categories, like odd and even.

In this chapter we look at the structure of numbers and different classifications of number.

> A **factor** is a number that divides *exactly* into another number without a remainder.

For example 5 is a factor of 15, because 5 divides exactly into 15 without a remainder (5×3=15).

However 6 is not a factor of 15, because 6 does not divide exactly. (There are two 6's in 15 and a remainder of 3).

Often we know the factors of a number by knowing our times tables and working backwards. In the example above we know our 5 times table and that 5 3's are 15, so 5 (and 3) are factors of 15.

1 Copy and complete the following table **up to the number 30**. Put **Table of Factors** at the top.

NUMBER	FACTORS	NUMBER OF FACTORS
1	1	1
2	1, 2	2
3	1, 3	2
4	1, 2, 4	3
5	1, 5	2
6	1, 2, 3, 6	4
7		
8		

2 Which number in your table has the fewest factors?

3 How many factors does it have?

4 Which numbers have the most factors?

5 How many factors do they have?

We see how different numbers are: 12, for example has many factors, whereas 13 which is only one more than 12 has only two.

SQUARE NUMBERS

You will need some counters.

* Start with one counter on the table:

* Add three more to form a larger square:

* Add five more to form a larger square:

6 Copy and complete the table below:

Number of Square	1	2	3	4	5	6
Number of Counters	1	4	9			

7 How many counters would you need for the 7th square?

8 How many counters would you need for the 11th square?

9 How many counters would you need for the 30th square?

10 Describe how the number of counters increases as the square size increases.

11 Do you see a connection between each square number and the number of counters needed to make it?

> The numbers you have found; **1, 4, 9, 16** and so on are called **square numbers**.

This is because you can arrange that number of counters to form a square. The 4 counters are in 2 rows of 2. The 9 counters are in 3 rows of 3 and so on.

We can also think of these numbers as being the answers to the following sums:

$$1 \times 1 = 1 \qquad 2 \times 2 = 4 \qquad 3 \times 3 = 9 \qquad 4 \times 4 = 16 \qquad 5 \times 5 = 25 \quad \text{etc.}$$

So, if we 'square a number' we multiply it by itself. 2 squared is 4; 3 squared is 9; 4 squared is 16 and so on. We can also write this as 2^2, 3^2, 4^2 etc. 2^2 means 2×2, $\mathbf{3^2 = 3 \times 3 = 9}$ and so on.

✷ Look at your table of factors completed earlier.

✷ Study your answers carefully to see what you notice. In particular, look at the square numbers and the number of factors column.

You may have noticed that square numbers all have an **odd** number of factors. 1 has only 1 factor; 4, 9 and 25 have 3 factors and 16 has 5 factors.

> Square numbers always have an odd number of factors.
> All other numbers have an even number of factors.

FACTOR PAIRS

✷ Look at your table of factors. You will see that 12 has 6 factors: 1, 2, 3, 4, 6, 12. If we pair them up like this:

1, 2, 3, 4, 6, 12

do you notice anything?

If you multiply the numbers in a pair you get 12: $1 \times 12 = 12$,
                                                       $2 \times 6 = 12$,
                                                       $3 \times 4 = 12$.

These pairs of numbers are known as **factor pairs.**

This pairing is useful, because if you know one factor of a number you can get another. For example, if you know 4 divides into 44 then another factor must be 11 because $4 \times 11 = 44$.

12 Write out the factors of **20** and show how they pair up.
13 Do this also for **18, 22, 24, 29**.

14 For each of the following numbers write down two pairs of factors: **32, 35, 40, 42, 45, 48, 55**.

FACTOR RECTANGLES

We can also represent factor pairs by rectangles. For example, the factor pairs of 12 can be represented in the following way:

OOOOOOOOOOOO   OOOOOO   OOOO
   $1 \times 12=12$ OOOOOO   OOOO
                         $2 \times 6=12$ OOOO
                                            $3 \times 4=12$

✳ Try making these with your counters and then draw them in your book.

15 Draw as many rectangles as you can for the number **20**.

16 Draw rectangles also for **18, 22, 24, 7, 17, 16**.

Since the factors of 16 are 1, 2, 4, 8, 16   we can pair up the 1 and 16, and the 2 and 8. But this leaves the 4. Clearly $4 \times 4=16$ and so we need to pair 4 with itself.

So you should have drawn three rectangles for the number 16 (A square is a special kind of rectangle).

This will apply to all square numbers since they always have an odd number of factors.

PRIME NUMBERS

You may have noticed that some of the numbers you were drawing rectangles for could only be drawn in one way: a single row of counters. The number 7, for example, can only be drawn as a row of 7 counters. It has only factors of 1 and 7.

Numbers which have exactly two different factors are known as **prime numbers.**
Their only factors are the number itself and one.

As we will see later prime numbers are very important.

The number 1 is not normally considered as being prime because according to the above definition prime numbers have exactly two different factors.

✳ Check that **2, 3, 5, 7, 11, 13, 17** are all prime numbers.

17 Write these numbers out and add all the other prime numbers up to 30 to the list.

18 Are there any even numbers in your list that are prime numbers (apart from the number 2)? Why not?

19 Which of the following are prime numbers: **31, 32, 36, 37, 39**.

Prime numbers have also been a source of great interest among mathematicians.
No one, for example, has yet been able to find a pattern in the sequence of prime numbers that could be used to find the following prime numbers.

THE SIEVE OF ERATOSTHENES

The Greek mathematician Eratosthenes found a neat way of deleting all non-prime numbers so that only the primes were left.

✳ Write out the numbers from 1 to 100: write 1 to 10 on the first line, then 11 to 20 underneath these on the next line and so on until you reach 100.

✳ The first prime number is 2, so cross out every 2nd number after 2 (4, 6, 8 etc.).
✳ The next prime number is 3, so cross out every 3rd number after 3 (6, 9, 12 etc.).
✳ The next prime number is 5, so cross out every 5th number after 5.
✳ Similarly cross out every 7th number after 7.

The next prime number is 11 but there is no need to go this far as no new numbers will be crossed out: the numbers not crossed out are the prime numbers from 1 to 100.

✳ Circle these with a colour.

TRIANGULAR NUMBERS

You will need some counters.

❊ Start with one counter on the table: ○   **1**

❊ Add two more to form a triangle: ○
                                            ○ ○   **2**

❊ Add three more to form a larger triangle: ○
                                                   ○ ○
                                                 ○ ○ ○   **3**

20 Copy and complete the table below:

Number of Triangle	1	2	3	4	5	6	7	8
Number of Counters	1	3	6					

21 How many counters would you need for the 9th triangle?
22 How many counters would you need for the 30th triangle?
23 Describe how the number of counters increases as the triangle size increases.
24 Do you see a connection between each triangle number and the number of counters needed to make it?

25 Write down 2 triangle numbers which are also square numbers.

CUBE NUMBERS

You will need some multilink cubes.

❊ Start with one cube on the table: ⬜ ¹

❊ Add 7 more to make a larger cube: ²

❊ Add 19 more to make a larger cube:

26 Copy and complete the table below:

Number of Cube	1	2	3	4	5	6
Number of Cubes	1	8	27			

27 How many cubes would you need for the 7th cube?
28 How many cubes would you need for the 10th cube?
29 How many cubes would you need for the 30th cube?
30 Describe the connection between the cube number and the number of cubes it is made from?

31 Which of your cube numbers are also square numbers?

The numbers you have found are called **cube numbers**.
This is because you can arrange that many cubes to form a larger *cube*.
The length, breadth and height of cubes are always the same.

We can also think of these numbers as being the answers to the following sums:
$$1 \times 1 \times 1 = 1$$
$$2 \times 2 \times 2 = 8$$
$$3 \times 3 \times 3 = 27$$
$$4 \times 4 \times 4 = 64 \quad \text{etc.}$$

So, if we *cube* a number, we multiply it by itself twice. 5 *cubed* is $5 \times 5 \times 5 = 125$.
We can also write this as 5^3.
5^3 means $5 \times 5 \times 5$.

This is similar to squaring a number where a number is multiplied by itself: 5^2 means 5×5.

32 What does 6^3 mean?
33 What does this equal?

34 What do you think 10^3 means?
35 What does this equal?

36 What do you think 6^4 means?   (we say '6 to the power 4')
37 What about 7^6?   (we say '7 to the power 6')

SUMMARY OF NUMBER SEQUENCES

The various sequences studied in this chapter will be referred to and used in later work and so they are listed here for easy reference.

In each case the sequence goes on forever.

Square numbers: **1, 4, 9, 16, 25, 36 . . . .**

Prime Numbers: **2, 3, 5, 7, 11, 13 . . . .**

Triangle Numbers: 1, 3, 6, 10 . . . .

Cube Numbers: **1, 8, 27, 64 . . . .**

"And there is a third nature, which is space, and is eternal ."

Plato (429-347 BC)

GREEK PHILOSOPHER

7

GEOMETRY

The third of the three branches of mathematics is geometry.
This deals with forms: lines, circles, solid objects and the way they are structured and compared.

All forms exist in space so all geometrical objects have some kind of space, whether it is length or space inside it or even just its position near some other object.

The simplest geometrical form is the point. All other forms are composed of points and so, in a way, the point contains them all.

By moving, a point can create a straight line or a circle and a circle can create a sphere (ball shape). More complex shapes can be made from simple shapes.

THE RIGHT ANGLE

Here are two straight lines meeting at a point.
One line goes straight up and the other goes across.

We call this a **Right Angle** because a square would fit exactly into it without leaving any gaps.

The walls of a room are built at right-angles to the floor.

A page of a book usually has a right angle in each corner.
The plus sign + shows four right angles.

Another way of saying that two lines are at right-angles to each other is to say that they are
perpendicular: the wall is perpendicular to the floor for example.

✴ List at least three other places where you can see right angles.

✴ Now make yourself a rectangle out of thin card or paper. The size is not important but make
   sure that the corners are as near to perfect right angles as you can make them. You may find it
   helpful to use a set-square to make it.

Next, in a group of 2 to 4 players you can play the Right-Angles Game.

☺ **RIGHT-ANGLES GAME**   A game for 2 to 4 players.

You may use a rectangular shape to test if an angle is right-angled.
Place the pack with the diagrams uppermost.
Throw a die to decide who goes first.
The first player decides how many right angles there are in the top card diagram and tells the
others.
If the answer is the same as that on the back of the card the player keeps the card, otherwise it
goes to the bottom of the pack.
The player on the first player's right goes next.
When all the cards have been taken the winner is the one with the most cards.

PARALLEL LINES

A pair of straight lines which do not meet each other even if they are extended as far as you like
are called **parallel** lines. Here are three examples:

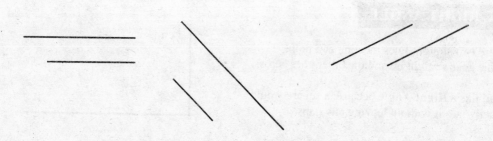

1 Can you think of a mathematical symbol which is a pair of parallel lines?

2 Name a mathematical shape which is made from two pairs of parallel lines at right-angles to each other.

DRAWING GEOMETRICAL SHAPES

Of course it is not possible to draw a perfect circle or triangle. A perfect triangle has three perfectly straight sides which have no thickness and which meet at three points. The shapes we draw should be as close as we can get them to the perfect shapes.

Look at Diagram A below which shows three circles.

Diagram A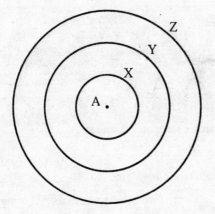

The circles are labelled X, Y and Z.
All the circles have the same centre, which is labelled A.

> The **radius** of a circle is a line from its centre to any point on the circle.
> The length of a radius is the distance from the centre of a circle to any point on the circle.

3 With a ruler measure the radius of circles X, Y and Z and write them down.

❋ Using your compasses copy the diagram into your book. Be as accurate as possible.

In the Diagram B below some points have been labelled. The circles are not labelled this time.

Diagram B

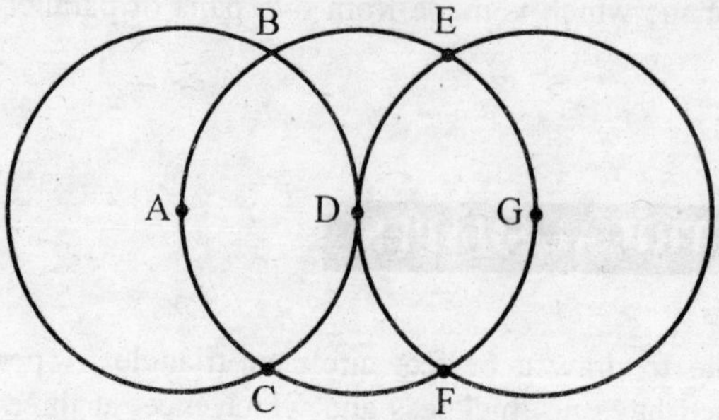

A is the centre of the circle on the left.
The circle on the left and the middle circle meet at B and also at C.

? **EXERCISE 1**

Copy and fill in:

a The centre of the middle circle is at ___.

b The centre of the circle on the right is at ___.

c The middle and right-hand circles meet at ___ and at ___.

d The left and right circles meet at ___.

e The length of the radius of each circle is ___ cm.

f The distance from A to G is ___ cm. (We can write this as <u>AG = 4cm</u>)

g BE = ___ cm.

h BD = ___ cm.

i BG = ___ cm.

✱ Now copy the diagram into your book exactly, using compasses.

With your ruler and a coloured pencil join C to F, join F to E and join E to C.

> The triangle CFE is a **right-angled triangle** because a
> right angle would fit into the triangle at F.

Diagram C below shows three circles P, Q and R.

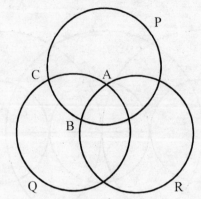

Diagram C

✳ Copy the diagram into your book. Copy all the letters as well.

✳ There are six points where the circles meet. Count them.

✳ Circles P and Q meet at C and D. Put D in the correct place on your diagram.

✳ Circles P and R meet at B and F. Mark F on your diagram.

✳ Q and R meet at E and A. Put E on your diagram.

4 Join points A, B and D in colour to make a triangle.

| ABD is an **equilateral triangle** because all its sides are equal in length. |

5 There are other equilateral triangles that can be drawn in the diagram.
For each one you can find write down the three letters that describe it.

RIGHT-ANGLED AND EQUILATERAL TRIANGLES

Diagram D

In Diagram D there are 3 small circles of radius 1.5cm.

✳ Copy these 3 circles exactly.

In the diagram there are also 3 large circles. Each of these has the same centre as a small circle and is twice its radius.

✳ Copy the 3 large circles onto your diagram.

✳ Draw in the straight line as well. Measure the length of this line. It should be 4.5cm.

✳ Now draw in colour on your diagram a right-angled triangle with a base of 4.5cm.

6 Draw on your diagram an equilateral triangle with sides 3cm.

> "*A* circle is a plane figure contained by one line such that all the straight lines falling upon it from one point among those lying within the figure are equal to one another."
>
> Euclid (definition) (c 300 BC)
>
> GREEK MATHEMATICIAN

8

SYMMETRY

Many forms in nature are symmetrical: flowers, the faces of people, the sun and moon and so on. There are two main types of symmetry: **line symmetry** and **point symmetry** (also called rotational symmetry). We look first at line symmetry.

A

✳ Draw the diagram opposite using compasses. A black dot is at the centre of each circle (the radius is 3cm).

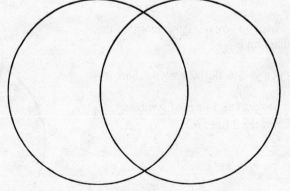

When lines meet we say that they **intersect** and the point where they meet is called the **point of intersection**.

✴ On your diagram put a dot at each of the two intersection points.

✴ Join your dots up like this:

This line is called a **line of symmetry** because if
you fold the paper along this line one half
would fold over onto the other half, the pattern
is **symmetrical**.

If you stand a mirror along the line you can see
that one half of the pattern is a **reflection** of the other.

Your diagram has another line of symmetry.
✴ Can you see it? It goes through the first pair of dots.

1 Draw the second line of symmetry in.

There are no other lines of symmetry.

B

✴ Draw the two circles again but without any dots or straight lines.

✴ At the upper intersection point put your
compass point and draw another circle
the same size as the others.

✴ Where do you think the centre of this
diagram is?

✴ Put a yellow dot where you think it is.

This pattern has 3 lines of symmetry.
2 Draw the 3 lines in.

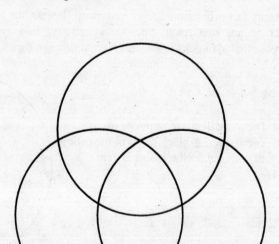

* Now draw the 3 circles again, but with a radius of 2cm. Leave some space around them.

* Put a red dot at the centre of each of the 3 circles.

There are 3 other intersection points.
* Can you see them?

* Put a blue dot at each of these 3 points.

* Draw 3 more circles the same size as before, centred on the blue dots.

How many new intersection points have appeared?
There should be 6.
* Put a red dot at each of these.

* Draw a circle centred on each of these red points.

You should have 9 new intersection points now.
* Put a blue dot on them.

* Using a ruler join these up in order with a colour.

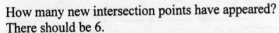

A 9-sided figure is called a **nonagon**.

* Write this word beside your diagram.
3 How many lines of symmetry does this nonagon have?
4 Draw them in.

POLYGONS

You will need Worksheet 1, which is an extended version of the pattern you have been drawing.
Write the title <u>Polygons</u> at the top.

You will be joining up intersection points in the pattern on the worksheet to make shapes.
You will need a ruler and some coloured pencils.

Can you see which intersection points to join up on the Worksheet
to make this shape?

triangle

✷ Join the points up using a colour.
In visualising a shape not actually there you are using the formula *By Mere Observation.*

5 Find the shapes below and join them up. Write the name of the shape as well.

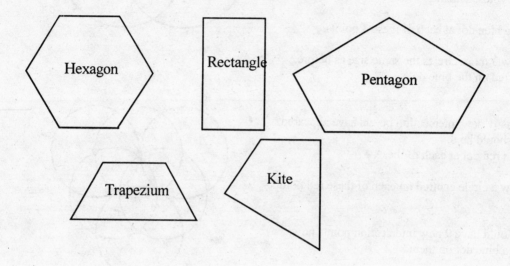

A 3-sided shape like the one you have drawn is called a **triangle.**
This particular triangle, you may remember, is an equilateral triangle (all sides are equal).

A 4-sided shape is called a **quadrilateral.**

6 3 of the shapes you have drawn are quadrilaterals. Which 3?

A 5-sided shape is called a **pentagon**.

A 6-sided shape is called a **hexagon**.
The particular hexagon you have drawn would be called a **regular** hexagon. The word regular tells
you that all the sides are equal in length.

7 Is the pentagon you drew a regular pentagon?

The shapes you have drawn are different types of polygons. A **polygon** is a many-sided figure.
✷ Copy into your book:- A triangle has 3 sides.
                        A quadrilateral has 4 sides.
                        A pentagon has 5 sides.
                        A hexagon has 6 sides.

        A regular polygon has all its sides of equal length.

ROTATIONAL SYMMETRY

Look at the three shapes below.

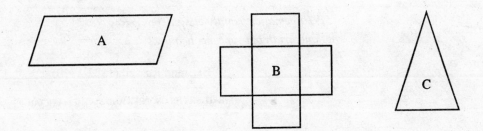

Shape A does not have any lines of symmetry, but it does have a kind of symmetry.
If this shape was rotated half a turn about its centre it would coincide with its original position.
So **2** half turns would bring it back to its original position.
We say that the shape has **rotational symmetry of order 2**.

Similarly if shape B is given a quarter turn it would coincide with its original position and **4** quarter turns returns the shape to its original position: it has rotational symmetry of order **4**.

Shape C requires **1** complete turn before returning to its original position so it has rotational symmetry of order **1**.

 EXERCISE 1

Copy the following shapes, put a dot at its centre (if it has a centre) and write down the order of its rotational symmetry:

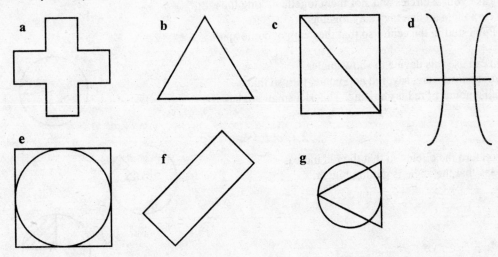

"Mathematics, rightly viewed, possesses not only truth but supreme beauty. . ."

Bertrand Russell (1872-1970)

BRITISH PHILOSOPHER AND MATHEMATICIAN

9

ANGLES AND TRIANGLES

✳ Draw a circle of radius 4cm on thin card.
Draw a radius somewhere.
Cut out the circle and also cut along the radius to the centre.

Make another identical circle but from a different colour card.
Take your 2 circles and slot them together along the radii.
Turn them so that they stay together.
Put a stud in the centre so that they cannot come apart.

You can use this device to show angles.
If your colours are, say, red and white, turn so that a
small amount of red is showing. This is a small angle.

If you turn the circles so that the red increases
we say that the angle is getting bigger.

This is a right angle, a quarter of a circle.

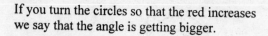

We can increase it further to a half-circle.
This is 2 right angles.

We can increase it to 3 right angles.

And then to a full circle, 4 right angles.

Angles smaller than a right angle are known as **acute** angles.

Angles between 1 and 2 right angles are called **obtuse**.

Angles between 2 and 4 right angles are called **reflex**.

✱ Write these words in your book together with the examples.

An angle measures the amount of turn.

A complete turn is usually divided into 360 degrees, written 360°.
This means that half a turn is 180°.
And a quarter of a turn, which is a right angle, is 90°.

So all acute angles are between 0 and 90 degrees.
All obtuse angles are between 90° and 180°.
And all reflex angles are between 180° and 360°.

? EXERCISE 1

Look at each diagram below and estimate the number of degrees in each angle marked with a letter.

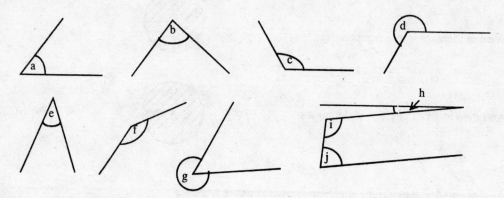

1 Which of the angles above are acute?
2 Which are obtuse angles?
3 Which are reflex angles?

 EXERCISE 2

Now you are given the size of the angle and you are asked to <u>sketch</u> the angle approximately (use pencil and ruler).

a 20°   **b** 100°   **c** 160°   **d** 70°   **e** 200°   **f** 260°   **g** 300°   **h** 45°

FINDING ANGLES

Knowing that 360° make a full turn we can find unknown angles.

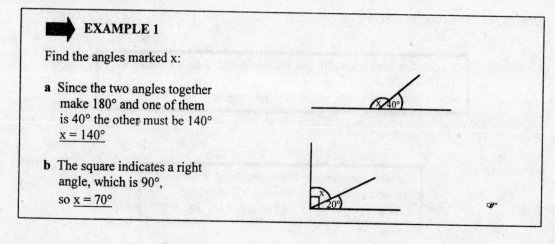

➡ **EXAMPLE 1**

Find the angles marked x:

a Since the two angles together
make 180° and one of them
is 40° the other must be 140°
x = 140°

b The square indicates a right
angle, which is 90°,
so x = 70°

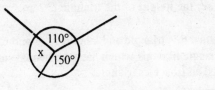

c The three angles make 360°
so *By Addition and by*
Subtraction we get x = 100°

 EXERCISE 3

Find x:

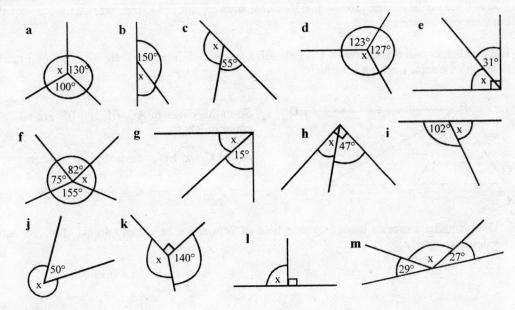

a   x 130° 100°

b   150° x

c   x 55°

d   123° 127° x

e   31° x

f   82° 75° x 155°

g   x 15°

h   x 47°

i   102° x

j   50° x

k   x 140°

l   x

m   29° x 27°

TRIANGLES

A 3-sided figure, a **triangle**, is extremely useful in mathematics.
The simplest triangle is the **equilateral triangle**
which has all its sides equal in length.
This triangle has 3 lines of symmetry.

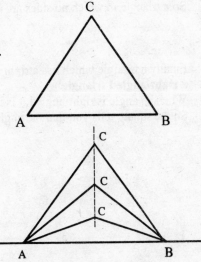

4 Copy the triangle and draw in the 3 lines of symmetry.
Write **equilateral triangle** beside it.

Now suppose that the base of the triangle is fixed
but the corner C is allowed to move up or down.
Wherever C is placed would you agree that the lines
AC and BC will be equal in length?

Suppose the base is 5cm long and that C is 2cm
above the base.

We say the **height** of the triangle is 2cm.

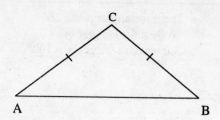

5 Draw this triangle and label it A, B and C.
A triangle like this which has just 2 equal sides is
called an **isosceles** triangle.

✳ Write **isosceles triangle** beside your diagram.
✳ Draw in the line of symmetry.

You will see on the diagram shown that the equal sides are marked to indicate that they are equal.
Put them on your triangle.

The equal sides will not always be marked with a single dash like this- there may be 2 or 3 marks:
sides with the same marking are equal.

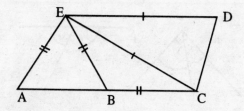

So in this diagram AE, BE and BC are all
equal in length.

Also EC and ED are equal to each other.

6 Draw another isosceles triangle with a base of 7cm and a height of 4 cm. Mark the equal
sides.

If a triangle is not equilateral or isosceles it is called **scalene**.
So a triangle in which no sides are equal is a scalene triangle.

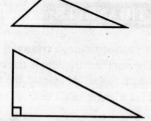

Finally a triangle which has a right angle in it is called
a right-angled triangle.
When an angle is right-angled it is usual to put a small square in it.
So the square tells you that the angle is 90°.

EXERCISE 4

For each triangle below say if it is

E,	Equilateral
I,	Isosceles
S,	Scalene
R,	Right-angled

(more than one of these will apply in some cases)

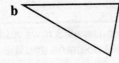 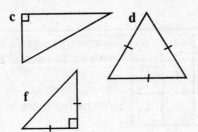 

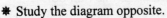

The diagram shows a figure ABEGC. It is described by its **vertices** (its corners) which are A, B, E, G and C.
You could say that
it is composed of a
rectangle BEGC and a triangle ABC.

Or you could say it is composed of 6 triangles.

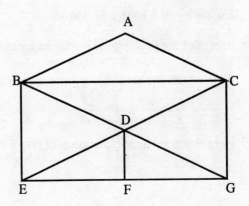

7 Describe the 6 triangles in terms of vertices.

8 Which line is parallel to BC?

9 Which 2 lines are parallel to DF?

10 Which line is parallel to EC?

✴ Study the diagram opposite.
✴ Measure the radius of a circle and draw the diagram carefully.

✴ Complete triangle ABC.

11 What kind of triangle is it?

12 Draw a line through C parallel to AB.

13 Draw in the rectangle BACD.

14 Find the point E which is on the line AB, and BE = 1cm? Mark the point E on your diagram.

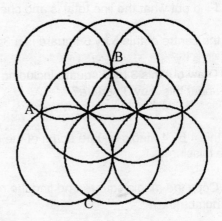

MAGIC SQUARES

A magic square is one in which the numbers in every row, column and each of the diagonals add up to the same total.

For example:

8	1	6
3	5	7
4	9	2

is a magic square because all 3 rows add up to 15, and so do each of the 3 columns and the 2 diagonals.

All the numbers from 1 to 9 are used.

✱ Check that you agree with this.

So there are 8 lines totalling 15.

In one of these lines the numbers increase by 1: this is the diagonal - 4, 5, 6.

1 In which line do the numbers increase by **a** 2, **b** 3, **c** 4?

Here is a 5 by 5 magic square (it has 5 rows and 5 columns).

23	1	2	20	19
22	16	9	14	4
5	11	13	15	21
8	12	17	10	18
7	25	24	6	3

2 Find out what the line total is and check that all 12 lines add up to this number.

In the centre of this 5 by 5 square is a 3 by 3 square.

3 Draw out this 3 by 3 square including the numbers in each box. What do you notice about this 3 by 3 square?

In this 4 by 4 magic square some of the numbers are missing.

16	3	2	
		11	8
9			12
4	15	14	1

4 Copy the magic square and find the missing numbers.

This is a 6 by 6 magic square.

32	31	1	3	21	23
29	30	4	2	24	22
9	11	20	19	25	27
12	10	17	18	28	26
16	15	33	35	5	7
13	14	36	34	8	6

5 Find the total and check that some rows and columns add up to this number.

The bold lines divide the square up into 9 parts.

6 What do you notice about the 4 numbers in each of these parts?

7 Add up the 4 numbers in each of the 9 boxes and put the results into a 3 by 3 square. (the top left square will have 32+31+30+29=122 in it)

8 What do you notice about your 3 by 3 square?

A **Magic Star** works in a similar way:

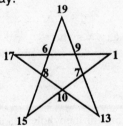

At every intersection of lines there is a number.

9 What do you find about the total of the numbers on every diagonal?

10 Construct a magic star like this but using the numbers 1, 2, 3, 4, 5 for the inside intersections and the numbers 5, 7, 8, 9, 11 for the outside ones (the total is 22).

"All the effects of nature are only the mathematical consequences of a small number of immutable laws."

Marquis de Pierre Simon Laplace (1749-1827)

FRENCH MATHEMATICIAN

10

BY THE COMPLETION OR NON-COMPLETION

Suppose you had to add the numbers   **19, 8, 1.**
✷ Is there an easy way to add them up?

It is easier to add 19 and 1 first, because it completes a 20.
So the answer would be **28**.

✷ Try **3 + 6 + 7 + 8 + 4.** What is the easy way here?
Well you could pair up 3 and 7 which complete a 10, and also 6 and 4 which complete another 10.
This gives the answer as **28** again.

> This technique of looking for the wholeness is described by the Sutra
> *By the Completion or Non-Completion.*

In the following exercise you may like to indicate the numbers being combined by joining them

up. For example 19+3+11+ 2 = **35**

 EXERCISE 1

Find an easy way of doing these sums:

a 29 + 7 +1 +5   **b** 16 + 3 + 6 + 17   **c** 8 + 51 + 12 + 3   ☞

d $37 + 7 + 21 + 13$ **e** $13 + 16 + 17 + 24$ **f** $73 + 46 + 54 + 8$

g $33 + 25 + 22 + 35$ **h** $18 + 13 + 14 + 23$

After some practice you may find you can quickly add a long string of numbers by this method.

 EXERCISE 2

Complete the following addition sums:

a $3 + 9 + 5 + 7 + 1$ **b** $27 + 15 + 23$

c $43 + 8 + 19 + 11$ **d** $32 + 15 + 8 + 4$

e $24 + 7 + 8 + 6 + 13$ **f** $6 + 33 + 24 + 17$

g $73 + 68 + 27$ **h** $41 + 16 + 23 + 9 + 14 + 3$

i $8 + 6 + 9 + 8 + 4 + 12 + 11 + 2$ **j** $17 + 12 + 13 + 9 + 21 + 18 + 9$

```
k  4 4        l  3 5        m  4 8        n  6 3 2 7        o  5 4 9
   2 2           7 6           3 8           5 8 4            1 8 2
   6 9           4 5 +         6 2           7 4 3 +          3 1 7
   8 6 +         ___           7 1 +         _____          2 4 1
   ___                         ___                            7 2 6
                                                              3 2 1 +
                                                              _____
```

COMPLETING THE WHOLE: FRACTIONS

When a whole one is divided we get fractions.
We can think of a whole one being divided into 2 halves or 3 thirds or 4 quarters

We can write this as:

$$1 = \tfrac{2}{2} \qquad 1 = \tfrac{3}{3}$$

Or geometrically:-

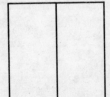 = 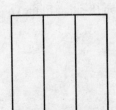 =

1 Copy out lines **(1)** and **(2)** above and continue the sequence up to the eighths.

Now suppose you have $\frac{3}{4}$ and you want to make it up to a whole one.

You will need $\frac{1}{4}$.
You can write $\frac{1}{4} + \frac{3}{4} = 1$.

EXERCISE 3

Write out these sums but replace the question mark with the correct fraction:

a $? + \frac{1}{2} = 1$ **b** $? + \frac{2}{5} = 1$ **c** $? + \frac{1}{3} = 1$ **d** $? + \frac{5}{6} = 1$

e $? + \frac{3}{8} = 1$ **f** $? + \frac{7}{10} = 1$ **g** $? + \frac{1}{7} = 1$ **h** $? + \frac{3}{11} = 1$

i $? + \frac{13}{20} = 1$ **j** $? + \frac{99}{100} = 1$

COMPLETING THE WHOLE: SHAPES

2 Here you are asked to solve three puzzles:

✳ Puzzle 1

Get the five pieces of Puzzle 1 and arrange them to form a dodecagon (12-sided figure).
Next rearrange them to form a square.

✳ Puzzle 2

Get the six pieces of Puzzle 2 and make an octagon (8-sided figure) out of them.
Then rearrange these pieces to form a square.

3

you will need to make the 5 pieces yourself as shown as follows:

On squared or graph paper draw a Greek Cross with all sides 4cm

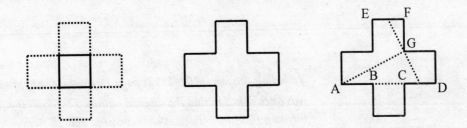

Draw the lines shown dashed above: B to C, A to G and the mid point of EF to the mid point of CD.

Then see if you can make each of the shapes shown below:

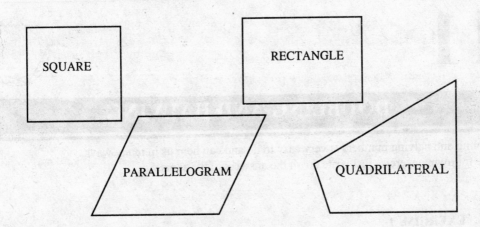

Now reconstruct the original Greek Cross

Jedediah Buxton (1702–1772) from Derbyshire, England was once asked to find the amount obtained by doubling a farthing 140 times. His answer consisted of a 39-figure number for the pounds plus 2s 6d. Asked to multiply this 39-figure number by itself he gave the answer after two months having calculated it mentally from time to time over that period.

11

DOUBLING AND HALVING

Doubling and halving numbers is very easy to do and can help us in many ways.
In the following exercise just write down the answers to the sums.

EXERCISE 1

Double the following numbers:

a 43	**b** 1234	**c** 17	**d** 71	**e** 77	**f** 707
g 95	**h** 59	**i** 38	**j** 55	**k** 610	**l** 383

Halve the following numbers:

m 64	**n** 820	**o** 36	**p** 52	**q** 94	**r** 126
s 234	**t** 416	**u** 38	**v** 156	**w** 456	**x** 57

Since $4 = 2 \times 2$ we can multiply a number by 4 by doubling it and doubling the answer.

EXAMPLE 1

Find 35×4

We simply double 35 to 70, then double 70 to 140.
So $35 \times 4 = 140$.

Similarly since $8 = 2 \times 2 \times 2$ we can multiply by 8 by doubling three times.

 EXAMPLE 2

Find 26×8

Doubling 26 gives 52, doubling 52 gives 104, doubling 104 gives 208.
So $26 \times 8 = 208$.

 EXERCISE 2

Multiply the following:

a 53×4 **b** 28×4 **c** 33×4 **d** 61×4

e 18×4 **f** 81×4 **g** 42×4 **h** 23×4

i 16×4 **j** 16×8 **k** 22×8 **l** 45×8

m 73×2 **n** 37×4 **o** 53×8 **p** 76×8

The next exercise involves multiplying some fractions.

 EXERCISE 3

Multiply the following:

a $8\frac{1}{2} \times 4$ **b** $11\frac{1}{2} \times 8$ **c** $19\frac{1}{2} \times 4$ **d** $2\frac{1}{4} \times 4$ **e** $5\frac{1}{2} \times 8$ **f** $9\frac{1}{2} \times 4$

g $7\frac{1}{2} \times 4$ **h** $16\frac{1}{2} \times 4$ **i** $3\frac{1}{2} \times 8$ **j** $30\frac{1}{2} \times 4$ **k** $3\frac{1}{4} \times 4$ **l** $9\frac{1}{4} \times 8$

The answers to these 12 sums can be converted to a special message according to the following scheme.

✳ Convert each answer to its digit sum and convert each digit sum to a letter using the following
code:

D	E	L	N	O	R	V	W	Y
1	2	3	4	5	6	7	8	9

Halving numbers is something which can also be repeated.
So if for example we halved a number and then halved again we would be
dividing the number by 4.

 EXAMPLE 3

Divide 72 by 4.

We halve 72 twice: half of 72 is 36, half of 36 is 18.
So $72 \div 4 = 18$.

 EXAMPLE 4

Divide 104 by 8.

Here we halve three times:
Half of 104 is 52, half of 52 is 26, half of 26 is 13.

So $104 \div 8 = 13$.

 EXERCISE 4

Divide by 4:

a 56   **b** 68   **c** 84   **d** 180   **e** 116   **f** 92   **g** 34

Divide by 8:

h 120   **i** 440   **j** 248   **k** 216   **l** 44

EXTENDING THE MULTIPLICATION TABLES

You may have noticed another way of doing some of the sums in Exercise 2.
For example, for 18×4 you may have thought that since you know that $9 \times 4 = 36$, then 18×4 must be double this, which is 72.

Similarly if you don't know 8×7 but you do know that $4 \times 7 = 28$, you can just double 28.
So $8 \times 7 = 56$.
The following examples assume you know your tables up to 10×10, but if you don't know all these you should still be able to find your way to the answer.

 EXAMPLE 5

Find 6×14.

Since we know that $6 \times 7 = 42$, it follows that $\underline{6 \times 14 = 84}$.

 EXAMPLE 6

Find 14×18.

Halving 14 and 18 gives 7 and 9, and since $7 \times 9 = 63$ we double this twice.
We get 126 and 252, so $\underline{14 \times 18 = 252}$.

 EXERCISE 5

Multiply the following:

extending multiplication

a 16×7 **b** 18×6 **c** 14×7 **d** 12×9

e 4×14 **f** 6×16 **g** 7×18 **h** 9×14

i 16×18 **j** 14×16 **k** 12×18 **l** 16×12

All this comes under the Vedic Sutra *Anurupyena*, which means *Proportionately*.
We will be seeing many other applications of this formula later on.

When Zerah Colburn, aged eight, was asked for the factors of 247,483 he replied 941 and 263; when asked for the factors of 171,395 he gave 5, 7, 59 and 83; asked for the factors of 36,083 he said there were none. Presented with 294, 967, 297 he gave the only factors: 641 and 6, 700, 417.

12

DIVISIBILITY

DIVISIBILITY BY 2, 5, 10

❋ Look at the following sequence of numbers:

$$2, \ 4, \ 6, \ 8, \ 10, \ 12, \ 14, \ 16......$$

1 What factor apart from 1 is common to all these numbers?
(Look at your table of factors from chapter 6 if necessary)
2 How can you tell if 2 is a factor of a number?
3 Does every even number have 2 as a factor?
4 Do any odd numbers have 2 as a factor?

> All numbers ending in an even number or zero must have 2 as a factor.
> We say the number is **divisible** by 2.
> We can also say that any number ending in an even number or zero is a **multiple** of 2.

So, given the number 38, we may say **2 is a factor of 38**
                                   or **38 is a multiple of 2**
                                   or **38 is divisible by 2.**

Look at the following sequence:

$$5, 10, 15, 20, 25, 30......$$

5 Which factor apart from one is common to all these numbers?

> **All numbers ending in 5 or 0 are divisible by 5.**

What about the following series of numbers?

$$10, 20, 30, 40, 50, 60....$$

6 What number apart from one, two or five divides into all of the above?

> **All numbers ending in 0 are divisible by 10.**

The Vedic formula which applies for testing for divisibility by 2, 5 and 10 is the sub-Sutra *Only the Last Terms.*

 EXERCISE 1

Which of the following numbers are divisible by 2 or 5 or 10?

a 12	**b** 17	**c** 15	**d** 20	**e** 28	**f** 70
g 61	**h** 124	**i** 145	**j** 16750	**k** 453628	**l** 1392392

 EXERCISE 2

Find a 2-figure number, divisible by:

a 2, whose figures add up to 5. (3 answers)
b 2, whose figures add up to 7. (4)
c 2, when the first figure is 3 times the last.
d 2, when the first figure is 6 less than the last.
e 2, when the product of the figures is 12. (3)
f 2, when the product of the figures is 2.
g 2, when the difference between the figures is 7. (3)
h 2, when the difference between the figures is 8.

i 2, when the last figure is 6 times the first.
j 2, when the difference between the figures is 3. (7)
k 5, if the product of the figures is 35.
l 5, if one figure is greater than the other by 7.

DIVISIBILITY BY 3 AND 9

Look again at your 9 point circle.

7 Write down all the numbers you put on your '9' branch. Write down what you think the next three numbers should be.

8 Do you notice anything? Where have you seen this set of numbers before?

> All numbers whose digit sum is 9 are divisible by 9.

Look again at your 9 point circle.

9 Write down all the numbers on your 3, 6 and 9 branches in order with the smallest first.

10 Do you notice anything? Where have you seen this set of numbers before?

> All numbers with a digit sum of 3, 6 or 9 are divisible by 3.

SUMMARY

We can summarise the divisibility tests for 2, 3, 5, 9, 10 as follows:

    For 2: is the last figure even or 0?
    For 3: is the digit sum 3, 6 or 9?
    For 5: is the last figure 5 or 0?
    For 9: is the digit sum 9?
    For 10: is the last figure 0?

If the answer is yes the number is divisible, otherwise it is not divisible.

DIVISIBILITY

11 Copy and complete the following table:

Number	Divisible by 2	Divisible by 3	Divisible by 5	Divisible by 9	Divisible by 10
12	yes	yes	no	no	no
15	no		yes		
18					
20					
30					
36					
45					
50					
12825					
645840					
71023					
663273					

 EXERCISE 3

Find a 2-figure number, divisible by:

a 3, if the first figure is 2. (3 answers)
b 3, if the first figure is 7. (3)
c 3, if the product of the figures is 14. (2)
d 3, when the first figure is 4, and the difference between the figures is 2.
e 3, when one figure is 5, and the difference is 4. (2)
f 3, when one figure is 8, and the other is odd. (4)
g 3, when both figures are even, and their difference is 0.
h 3, when the first figure is odd and their sum is 6. (3)
i 3, when the last figure is 3 times the first.
j 3, when the last figure is half the first. (4)
k 3, when the last figure is 4. (3)
l 3, if both figures are even. (5)
m 5, if the difference between the figures is 4. (3)
n 5, if both figures are odd, and the first is greater than the last. (2)
o 5, if both figures are odd and their difference is 4. (2)
p 5, if only one figure is odd, and their difference is 3. (3)
q 2, when the last figure is half the first. (2)
r 2, when the product of the figures is 28.

s 9, if the first figure is twice the second.
t 9, if both figures are prime.   (2)
u 10, if the digit sum is 7.
v 6 and 10.   (3)

DIVISIBILITY BY 4

We saw earlier that we can tell if a number can be divided exactly by 2 by looking at the last figure—if it is even or 0 then it can be divided by 2.

This works because any number is a multiple of ten plus its last figure.

For example   **34 = 30 + 4,   543 = 540 + 3.**

And since 2 will always divide into the first part (30 and 540 above) the full number will be divisible by 2 if the last figure is. So 34 is divisible by 2 and 543 is not.

We can extend this idea:

> If 4 divides into the last **two** figures of a number then 4 divides into the whole number.

 EXAMPLE 1

In the number **1234** we can ignore the 12 at the beginning and see if 4 divides into 34. We can do this by halving 34 twice: half of 34 is 17, and half of 17 is not a whole number, so we can say that <u>234 is not divisible by 4.</u>

 EXAMPLE 2

Is **3456** divisible by 4?

Just look at the last two figures, 56. Can 56 be halved twice without going into fractions?

Half of 56 is 28 and half of 28 is 14, so the answer is yes,
<u>3456 can be divided by 4.</u>

The reason this method works is that **3456 = 3400 + 56** and 4 must divide into 3400 (because 4 divides into 100) so we only need to find out if 4 divides into the last two figures.

 EXERCISE 4

Test the following numbers for divisibility by 4 (just write yes or no):

a 48084   **b** 468   **c** 112233   **d** 7007   **e** 6666   **f** 99008

g 7654   **h** 74   **i** 82182   **j** 55000   **k** 19181716   **l** 949596

This method comes under the Vedic Sutra *Anurupyena* which means *Proportionately*, but another Sutra can be used to do the same job.

> This is *Sopantyadvayamantyam* which means *The Ultimate and Twice the Penultimate*. The ultimate means the last figure, and the penultimate is the figure before the last one.

 EXAMPLE 3

So in the number **12384** the Sutra tells us to add up the 4 and twice the 8.
This gives us 20, and since <u>4 goes into 20 it will also go exactly into 12384</u>.

> So when using *The Ultimate and Twice the Penultimate* we add the last figure to twice the one before it,
> and if 4 divides into the result then the number is divisible by 4.
> Otherwise it is not divisible by 4.

 EXAMPLE 4

In the number **5574** the Sutra gives us 4 plus twice 7, which is 18.
But 4 will not divide exactly into 18 so <u>5574 is not divisible by 4</u>.

 EXERCISE 5

Look again at Exercise 4 and use *The Ultimate and Twice the Penultimate* to test the numbers again for divisibility by 4.
Write down the totals the Sutra gives you and then write down whether 4 divides into the number or not.

So far we have dealt with divisibility by 2, 3, 4 and 5. What about divisibility by 6?

DIVISIBILITY BY 6

Any number which is divisible by 6 must also be divisible by 2 and by 3 because 6 = 2 x 3.
This means that to test a number for divisibility by 6 it must pass the tests for **both** 2 and 3.

> All numbers divisible by **both 2 and 3** are divisible by 6.

EXAMPLE 5

78 is divisible by 2 (as it ends in an even number) and also by 3 (as its digit sum is 6). So 78 is divisible by 6.

 EXAMPLE 6

However **454** is divisible by 2, but not by 3, so it is not divisible by 6.

It is essential that **both** tests are passed.

 EXERCISE 6

Test the following numbers for divisibility by 6:

a 888   **b** 6789   **c** 3456   **d** 94   **e** 1234   **f** 2468012

g 1998   **h** 390   **i** 44444   **j** 70407   **k** 1110222   **l** 12345678

DIVISIBILITY BY 15

This is similar to divisibility by 6.

Because $15 = 3 \times 5$ any number which is divisible by 3 and by 5 must be divisible by 15.

 EXAMPLE 7

Consider the number **345**.
It must be divisible by 3 as its digit sum is 3, and it is also divisible by 5 as the last figure is 5. So 345 is divisible by 15.

 EXAMPLE 8

But the number **4040** is not divisible by 15 because although it is certainly divisible by 5 it is not divisible by 3.

Again both tests must be passed.

 EXERCISE 7

Test these numbers for divisibility by 15:

a 78   **b** 785   **c** 1785   **d** 88888   **e** 12345   **f** 543210

Here is a summary of all the divisibility tests we have had so far:

Number Being Tested	Test
2	Is the last figure 0 or even?
3	Is the digit sum 3, 6 or 9?
4	Is the 2-figure number on the end divisible by 4?
5	Is the last figure 0 or 5?
6	Is the number divisible by both 2 and 3?
9	Is the digit sum 9?
10	Is the last figure 0?
15	Is the number divisible by both 3 and 5?

 EXERCISE 8

For each of the following numbers test for divisibility by 2, 3, 4, 5, 6, 9, 10, 15:

a 525   **b** 45   **c** 864   **d** 57330   **e** 88188   **f** 181

g 842   **h** 999   **i** 121314   **j** 906030   **k** 9876   **l** 123456

m What is the smallest 2-figure number that does not pass any of these tests?

n What is the largest 2-figure number that does not pass any of these tests?

o What is the smallest number that passes all of these tests?

 EXERCISE 9

Find a 2-figure number divisible by:

a 4 if the first figure is a 7   (2 answers)

b 4 if the sum of the figures is 5

c 4 if the product of the figures is 8

d 4 if the last figure is one more than the first   (2)

e 4 if the first figure is odd and the last is 6   (5)

f 4 if the figures differ by 3   (3)

g 6 if the first figure is 4   (2)

h 6 if the first figure is greater than 7 and is odd   (2)

i 6 if the product of the figures is 56

j 6 if the first figure is twice the last   (2)

k 6 if the last figure is 4 more than the first

l 6 if twice the product of the figures is 36

? EXERCISE 10

Find a 3-figure number divisible by:

a 9 if the first and last figures are both 1

b 9 if the last figure is 6 more than the first   (3)

c 9 if the first two figures are the same and both even   (4)

d 9 if the first figure is 5 and the others are both odd   (2)

e 9 if all three figures are the same   (3)

f 9 and nearest to 200

g 15 if the first figure is 1 and all the figures are odd   (2)

h 15 if the first figure is 8 and the others are odd

i 15 and is nearest to 700

j 15 which is the same when written backwards   (3)

k 15 which is less than 200 and is also an even number   (3)

l 15 which begins with an even number and contains exactly two odd figures   (?)

m 15 with all its figures odd

n Find a 4-figure number divisible by 6 in which the figures are consecutive and   **i** ascending
                                                                                                                                                    ii descending

o Find the largest odd 3-figure number in which the figures add up to 7

p Find the smallest 3-figure number divisible by 5 in which the figures add up to 10

13

SHORT MULTIPLICATION AND DIVISION

MULTIPLICATION

Multiplying numbers, for example 263 × 3, is a straightforward process. We set the sum out as shown below, and multiply each figure in 263 by 3, starting at the right:

 EXAMPLE 1

Sum:	2 6 3	Check: 2
	3 ×	3 ×
	7 8 9	6
	1	

✳ Check that you agree with the answer 789.

The digit sum check has also been carried out above. The digit sums of the numbers being multiplied are 2 and 3, and when these are multiplied we get 6. Since the digit sum of the answer, 789, is also 6 this shows us that the answer is probably correct.

 EXAMPLE 2

6 2	check:	8
4 ×		4 ×
2 4 8		5 (since 8×4=32 and 3+2=5)

✳ Check that you agree that the check here confirms the answer, since the digit sum of 248 is the same as the digit sum of 8×4.

 EXERCISE 1

Multiply the following numbers and check each one using the digit sums:

a 83 × 4   **b** 67 × 3   **c** 23 × 4   **d** 18 × 3   **e** 33 × 4   **f** 55 × 8

g 77 × 3   **h** 79 × 4   **i** 88 × 8   **j** 234 × 5   **k** 717 × 6   **l** 32 × 3

m 73 × 4   **n** 533 × 2   **o** 3115 × 3   **p** 314151 × 4   **q** 19 × 8   **r** 343536 × 3

s 432101234 × 4   **t** 142857×7   **u** 371 × 9   **v** 6125 × 9

w 12345679 × 9

You may recall from Chapter 5 that the digit sums of the numbers in the 9-times table were always 9. In fact when any number is multiplied by 9 the digit sum of the answer is 9, and this is confirmed in the last three questions above.

1 Study the following pattern of numbers, check that each line is correct, and add three more rows:

$$0 \times 9 = 1 - 1$$
$$1 \times 9 = 11 - 2$$
$$12 \times 9 = 111 - 3$$
$$123 \times 9 = 1111 - 4$$

MULTIPLICATION BY 11

 EXAMPLE 3

Find **a** 52 × 11 and **b** 57 × 11.

The 11 times table is easy to remember and multiplying longer numbers by 11 is also easy.
a To multiply a 2-figure number, like 52, by 11 we write down the number being multiplied, but put the total of the figures between the two figures: 572.
So 52 × 11 = <u>572</u> between the 5 and 2 we put 7, which is 5+2.

b For 57 × 11 there is a carry because 5+7=12 which is a 2-figure number.
So we get 5₁27 = <u>627</u> the 1 in the 12 is carried over to the 5 to give 6.

This is easy to understand because if we want, say, 52×11 we want eleven 52's.
This means we want ten 52's and one 52 or 520 + 52:   5 2 0
$$\frac{5\ 2\ +}{5\ 7\ 2}$$ note how the 2 and the 5 get added

in the middle column.

 EXERCISE 2

Multiply the following by 11:

a 23   **b** 61   **c** 44   **d** 16   **e** 36   **f** 50   **g** 76

h 88   **i** 73   **j** 65   **k** 75   **l** 99   **m** Is 473 divisible by 11?

 EXAMPLE 4

Find 234 × 11.

3-figure numbers like this are also easy to multiply by 11.
234 × 11 = 2574 the first and last figures are the same as in 234.
For the 5 in the answer we add up 2 and 3 and for the 7 we add up 3 and 4.

 EXAMPLE 5

Find 777 × 11.

The method above gives: 7_14_147 = 8547. We simply carry the 1's over.

 EXERCISE 3

Multiply by 11:

a 345   **b** 444   **c** 135   **d** 531   **e** 888   **f** 372   **g** 629

How could this method be extended for multiplying numbers like 2345?

DIVISION

 EXAMPLE 6

Find 30 ÷ 7.

$$7\overline{)3\ 0}$$
$$4\ \ remainder\ 2$$ So 30 ÷ 7 = 4 rem 2.

One way of checking our answer is to multiply 4 by 7 and add the remainder on:
4 × 7 + 2 = 30. Since we get the 30 back the answer is correct.

EXERCISE 4

Divide the following and check them as shown in the example above:

a 6)2 5   **b** 3)2 9   **c** 4)3 5   **d** 8)4 5

e 7)5 1   **f** 3)1 2 2   **g** 5)1 5 4   **h** 4)8 0 9

 EXAMPLE 7

Find 3456 ÷ 7.

$$7)3\ 4^6\ 5^2\ 6$$
$$\underline{4\ \ \ 9\ \ \ 3}\ \ \text{remainder 5}$$

The division is done in the usual way: 34÷7 = 4 rem 6, placed as shown,
                                          65÷7 = 9 rem 2, placed as shown,
                                          26÷7 = 3 rem 5, as shown.

So 3456 ÷ 7 = 493 rem 5.

THE DIGIT SUM CHECK FOR DIVISION

The same check as above can be applied: 493 × 7 + 5 = 3456 but multiplying 493 by 7 takes a
little time and we may prefer to use the digit sum check instead: we want to show that
493×7 + 5 = 3456   is correct and we do this by changing each number to its digit sum:
493 has a digit sum of 7, 3456 has a digit sum of 9,
so **493 × 7 + 5 = 3456** becomes **7 × 7 + 5 = 9**
and this is true because 7×7=49=4, and 4+5 = 9.

 EXAMPLE 8

Find 70809 ÷ 6.

$$6)7\ ^1 0\ ^4 8\ ^0 0\ ^0 9$$
$$\underline{1\ \ 1\ \ 8\ \ 0\ \ 1}\ \ \text{rem 3}$$ this is the answer and for the

check we show that 11801×6 + 3 = 70809 is true in digit sums.
This becomes 2×6 + 3 = 6 in digit sums and it is correct
since 2×6=3 in digit sums and 3 + 3 = 6.

 EXERCISE 5

Divide the following and check using the digit sums:

a 3)4 6 8 1   **b** 4)9 1 3   **c** 5)7 0 3 1   **d** 6)3 2 1

e 7)2 2 2   **f** 8)9 0 8 0   **g** 9)1 0 0 1   **h** 2)3 4 5 6 7

DIVISION BY 9

As we have seen before, the number 9 is special and there is a very easy way to divide by 9.

 EXAMPLE 9

Find 23 ÷ 9.

The first figure of 23 is the answer: 2.
And we add the figures of 23 to get the remainder: 2 + 3 = 5.
So 23 ÷ 9 = 2 remainder 5.

It is easy to see why this works because every 10 contains a 9 with 1 left over.
So 2 tens contains 2 9's with 2 left over.
And if 20 contains 2 9's remainder 2, then 23 (which is 3 more) contains 2 9's remainder 5.

 EXERCISE 6

Divide by 9:
a 51   **b** 34   **c** 17   **d** 44   **e** 60   **f** 71   **g** 46

This can be extended to the division of longer numbers.

 EXAMPLE 10

Find **2301 ÷ 9**.

The sum can be set out like this: 9) 2 3 0 1

The 2 at the beginning of 2301 is brought   9) 2 3 0 1
straight down into the answer: ↓
                                                       2

This 2 is then added to the 3 in 2301, and 5 is put down: 9) 2 3 0 1
                                                                      ↗
                                                                   2 5

This 5 is then added to the 0 in 2301, and 5 is put down: 9) 2 3 0 1
                                                                      ↗
                                                                   2 5 5

This 5 is then added to 1 to give the remainder, 6:   9) 2 3 0 1
                                                                ↗
                                                             2 5 5 r6

The first figure of the number being divided is the first figure of the answer, and each figure in the answer is added to the next figure in the number being divided to give the next figure of the answer.
The last number we write down is the remainder.

 EXAMPLE 11

Find $1234 \div 9$.

$$9 \overline{)\ 1\ \ 2\ \ 3\ \ 4}$$
$$1\ \ 3\ \ 6\ r\ 10$$

In this example we get a remainder of 10, and since this contains another 9 we add 1 to 136 and get <u>137 remainder 1</u>.

 EXERCISE 7

Divide the following numbers by 9:

a 212	b 3102	c 11202	d 31	e 53
f 203010	g 70	h 114	i 20002	j 311101
k 46	l 234	m 56	n 444	o 713

 EXAMPLE 12

$3172 \div 9$

$$9 \overline{)\ 3\ \ \ 1\ \ \ 7\ \ \ 2}$$
$$3\ \ \ 4\ \ 11\ \ r\ 13$$

Here we find we get an 11 and a 13: the first 1 in the 11 must be carried over to the 4, giving 351, and there is also another 1 in the remainder so we get <u>352 remainder 4</u>.

 EXERCISE 8

Divide the following by 9:

a 6153	b 3282	c 555	d 8252
e 661	f 4741	g 12345	h 4747

14

POWERS OF TEN AND DECIMALS

You will already be familiar with decimals: numbers like 1234.5 which have a decimal point in them.
You will know that the 1 in this number represents 1000 as it is the thousands position (the fourth position before the decimal point)

Similarly the 2 means 200, the 3 means 30 and the 4 means just 4.

The decimal point shows where the whole number part of the number ends. Past the decimal point we go into fractions: first tenths then hundredths, thousandths and so on.

. . . hundreds tens units **tenths hundredths thousandths**   **ten   hundred**
                                                                 thousandths thousandths . . . .

1 What would you say comes after the hundred thousandths?

So the 5 in 1234.5 represents 5 tenths.

Every whole number has a decimal point, even though it may not be written.
So in the number 20 for example there is a decimal point at the end (20.) but it is not written as there is no need to write it.

ADDING AND SUBTRACTING DECIMALS NUMBERS

Since we should always add like things, and not unlike things:

> When adding and subtracting decimal numbers we should always line up the decimal points, then all the columns will be lined up: tens under tens, units under units and so on.

 EXAMPLE 1

Find   **a**   45.67 + 123.3   **b**   1.045 + 33 + 0.3

     a   4 5 . 6 7   **b**   1 . 0 4 5
        1 2 3 . 3   +   3 3
       1 6 8 . 9 7   0 . 3   +
                              3 4 . 3 4 5

The decimal points are simply lined up, remembering that in the number 33 the point is at the end.

 EXAMPLE 2

Find   **a**   76 – 1.23   **b**   5 – 0.005

     a   7 6 . 0 0   **b**   5 . 0 0 0
         1 . 2 3   –   0 . 0 0 5   –
       7 4 . 7 7   4 . 9 9 5

You may prefer to put the point at the end of whole numbers when subtracting, and as many zeros as necessary.

> **The Digit Sum Check** can be used in all our decimal work, just as in sums without decimal points.

EXERCISE 1

Do the following decimal additions and subtractions:

a 43.22 + 7.7   **b** 103.44 + 18   **c** 1.006 + 321.4   **d** 87 + 1.23

e 1.9 + 11.1 + 4   **f** 44.44 + 5.05 + 1   **g** 109.8 + 77.88   **h** 17.1 + 1.71 + 0.07

i 44.55 – 2.3   **j** 43.2 – 7.15   **k** 91.19 – 3.3   **l** 33 – 1.4

m 67.8 – 19   **n** 0.66 – 0.05   **o** 1.111 – 0.08   **p** 51.5 – 3.57

MULTIPLYING AND DIVIDING DECIMAL NUMBERS

This is similar to ordinary multiplication and division.

> We simply insert the decimal point when we come to it.

 EXAMPLE 3

Find **a** 23.45×6 **b** $23.45 \div 2$

$$
\begin{array}{r}
\mathbf{a}\quad 2\,3.\,4\,5 \\
6\ \times \\
\hline
1\,4\,0.\,7\,0 \\
{}_{2}\ {}_{2}\ {}_{3}
\end{array}
$$

We multiply 6 by 5 and then by 4, as usual, then put the point down and continue. The final answer can be given as <u>140.7</u> since the final 0 serves no purpose.

$$
\mathbf{b}\quad 2)\overline{2\,3.\,{}^{1}4\,5\,{}^{1}0} \\
\underline{1\,1.\,7\,2\,5}
$$

We divide from the left, putting the point down when we come to it.
The extra 0 at the end of 23.45 has been put on so that the division can be completed.

 EXERCISE 2

Multiply/divide the following:

a 3.45×3 **b** 19.2×4 **c** 0.005×7 **d** 0.08×5 **e** 0.6×7 **f** 0.09×9

g $3.45 \div 3$ **h** $111.6 \div 9$ **i** $123.45 \div 5$ **j** $1.2 \div 6$ **k** $8.4 \div 7$ **l** $3.6 \div 5$

MULTIPLYING AND DIVIDING BY POWERS OF TEN

Take any number, say **3**, and multiply it by 10.
The answer is **30**.

Now divide **30** by 10.
The answer is **3**. We come back to the original number, 3.

> Any whole number can be multiplied by 10 by simply adding a 0 at the right-hand end.
> And any whole number that ends in 0 can be divided by 10 by simply removing that 0.

 EXAMPLE 4

So $55 \times 10 = \underline{550}$, $8760 \times 10 = \underline{87600}$,
and $390 \div 10 = \underline{39}$, $7000 \div 10 = \underline{700}$.

Since multiplying by 10 means adding a 0, multiplying by 10 **twice** means adding two 0's.
So to multiply by 100 (10 × 10) we put two 0's at the end.
Multiplying by 1000 we put three 0's, and so on.

Similarly for division: dividing by 100 involves removing two 0's,
dividing by 1000 means removing three 0's, and so on.

 EXAMPLE 5

So   66 × 100 = 6600,   230 × 1000 = 230000,
and   23400 ÷ 100 = 234,   9000000 ÷ 10000 = 900.

These sums are so easy that they come under the Vedic Sutra *Vilokanam* which means *By Mere Observation.*

 EXERCISE 3

Multiply/divide the following:

a 73 × 100   b 300 × 10   c 16 × 1000   d 470 × 10000

e 500 ÷ 10   f 91000 ÷ 100   g 3000 ÷ 1000   h 3003000 ÷ 100

Now we can bring in the *Proportionately* Sutra.
If we want to multiply a number by 20, which is 2 × 10, we need to double the number **and also** put a 0 on the end.

Multiplying by 300, we would multiply by 3 and put two 0's on the end.

 EXAMPLE 6

22 × 20 = 440,   300 × 22 = 6600,   80 × 40 = 3200.

With a sum like 80 × 40 above it is best to ignore the 0's, multiply 8 by 4 to get 32, and then put 2 0's on the 32, because there are 2 0's in the sum.

 EXAMPLE 7

So 60 × 30 = 1800,   300 × 20 = 6000   and   90 × 4000 = 360000.

 EXERCISE 4

Multiply the following:

a 333 × 20   b 45 × 20   c 200 × 44   d 26 × 2000   e 60 × 200   f 40 × 30

g 800 × 90   **h** 460 × 200   **i** 130 × 30   **j** 70 × 70   **k** 9000 × 9000   **l** 2320 × 300

m how many days in 400 weeks?   **n** how many minutes in 30 hours?

o how many hours in 20 days?   **p** if there are 30 weeks of school in a year, how many weeks of school are there in 12 years?

Now let us look at dividing by numbers like 20, 300 etc..
For dividing by 20 we need to divide by 2 and by 10.
So this means halving the number and taking a 0 from the end.

 EXAMPLE 8

a 660 ÷ 20 = <u>33</u>,   **b** 66000 ÷ 200 = <u>330</u> (we take two 0's from the 66000 here),
c 9000 ÷ 30 = <u>300</u>,   **d** 2460000 ÷ 200 = <u>12300</u>.

 EXERCISE 5

Divide the following:

a 6000 ÷ 20   **b** 8000 ÷ 400   **c** 1200 ÷ 60   **d** 210 ÷ 70

e 63000 ÷ 700   **f** 8800 ÷ 200   **g** 60 ÷ 20   **h** 963000 ÷ 3000

MULTIPLYING AND DIVIDING DECIMALS BY 10, 100 ETC.

When we multiply or divide a whole number by 10, 100 etc. we are just adding or removing 0's, so we are effectively just moving the decimal point.

Multiplying by 10, 100 , 1000 etc. means moving the point 1, 2, 3 etc. places to the right.
Dividing by 10, 100, 1000 etc. means moving the point 1, 2, 3 etc. places to the left.

 EXAMPLE 9

a 55.55 × 10 = <u>555.5</u>   **b** 1.234 × 100 = <u>123.4</u>   **c** 0.056 × 100 = <u>5.6</u>,

 EXAMPLE 10

a 3.69 × 100 = <u>369</u>   **b** 3.69 × 1000 = <u>3690</u>   **c** 0.7 × 100 = <u>70</u>,

Here we see that if the point is moved to the end of the number we do not need to write the point. And if the point has to go further than the end of the number a zero should be added or as many zeros as are needed to move the point the required number of places.

EXAMPLE 11

a $2.3 \times 20 = \underline{46}$ **b** $0.8 \times 300 = \underline{240}$.

Here we must move the point and also multiply the number.

 EXERCISE 6

Multiply the following:

a 55.56×10 **b** 12.345×10 **c** 3.7×30 **d** 7.177×100

e 0.88×10 **f** 1.232×2000 **g** 6.7×10 **h** 22.333×300

i 0.9×10 **j** 0.9×100 **k** 0.1×1000 **l** 8.76×100

m 1.3×700 **n** 4.44×10000 **o** 0.005×1000 **p** 0.07×30000

 EXAMPLE 12

a $55.55 \div 10 = \underline{5.555}$ **b** $1.23 \div 10 = \underline{0.123}$ **c** $15.51 \div 100 = \underline{0.1551}$,

 EXAMPLE 13

a $0.123 \div 10 = \underline{0.0123}$ **b** $7.7 \div 100 = \underline{0.077}$ **c** $76 \div 1000 = \underline{0.076}$,

 EXAMPLE 14

a $64.2 \div 20 = \underline{3.21}$ **b** $18.6 \div 30 = \underline{0.62}$ **c** $0.8 \div 400 = \underline{0.002}$.

Here we must move the point to the left and also divide the number (by 2, 3, 4).

 EXERCISE 7

Divide the following :-

a $333.3 \div 10$ **b** $333.3 \div 100$ **c** $444.4 \div 1000$ **d** $0.77 \div 10$ **e** $57.9 \div 10$

f $16.3 \div 100$ **g** $8.8 \div 100$ **h** $0.2 \div 1000$ **i** $1234.5 \div 1000$ **j** $1.2345 \div 1000$

k $0.05 \div 10$ **l** $3.04 \div 100$ **m** $83 \div 100$ **n** $300 \div 1000$ **o** $88.88 \div 20$

p $186.4 \div 200$ **q** $90.6 \div 30$ **r** $71.4 \div 70$

EXERCISE 8

This exercise is a mixture of the various types studied in this chapter.

a 40×70 **b** 32×100 **c** 640×10 **d** $640 \div 10$ **e** 20×100 **f** 30×50

g 22×40 **h** 34.56×10 **i** 0.5×100 **j** $4000 \div 20$ **k** $737.3 \div 100$ **l** 300×800

m 32×300 **n** 1.23×20 **o** 6.3×100 **p** $6.3 \div 100$ **q** 0.03×300 **r** 2.002×10

s $400 \div 10$ **t** $600 \div 30$ **u** $21.3 \div 30$ **v** $24.68 \div 200$ **w** $4.23 \div 30$ **x** $2.4 \div 400$

METRIC UNITS

Nowadays, most units we use are given using the **metric system**. There are certain standard units of measurement against which everything else is measured.

For example, mass is measured in kilograms. The **kilogram** is the mass of the International Prototype kilogram (a platinum cylinder) kept in a vault beneath the streets of Paris at the Institute of Weights and Measures.

Similarly, the **metre** is the distance between two specified marks on a specially shaped bar of platinum-iridium alloy, also kept at the Institute of Weights and measures in Paris. (The metre was originally determined as one 10 millionth of the distance between the equator and the pole).

Also since 1964, the **litre** is the standard measure for capacity (or volume of liquids).

The standard units of measurement are as follows (abbreviations are given in brackets):

> Distance = metre (m)   Mass = gram (g)   Capacity = litre (l)

A convenient feature of the metric system is the fact that it uses the decimal system for larger and smaller quantities, which means to convert one unit into another, we normally only have to multiply or divide by powers of ten.

The French words 'cent', 100 and 'mille', 1000, are used for fractions of the above units.
And 'kilo' is used for multiples of 1000.

The most commonly used units are:

> **CAPACITY**
> Basic unit of measurement = litre (l).
> There are 1000 millilitres (ml) in a litre.

MASS
Basic unit of measurement = gram (g)
There are 1000 milligrams (mg) in a gram
There are 1000 grams in a kilogram (kg)
There are 1000 kilograms in a metric tonne (t)

LENGTH
Basic unit of measurement = metre (m)
There are 1000 millimetres (mm) in a metre   (Note: There are 10 mm in a centimetre)
There are 100 centimetres (cm) in a metre
There are 1000 metres in a kilometre (km)

It is often important to be able to change one unit into another. For example, we may calculate the answer to a sum as being say 789,000 metres and it may be more concise and neater to give the answer as 789 km. Similarly, it may be better to give an answer of 0.025 kg as 25 grams.

 EXAMPLE 15

Convert 27,000 metres into kilometre.

There are 1000 metres in a kilometre, so we divide by 1000:
27,000 ÷ 1000 = <u>27km</u>.

 EXAMPLE 16

Change 0.07 litres into millilitres.

There are 1000 ml in a litre, so 0.07 litres must be 0.07 × 1000 = <u>70 ml</u>

 EXERCISE 9

Change the following quantities into the required units:

a 36,000 mm into centimetres
b 36,000 mm into metres
c 36,000 cm into metres
d 234,000 m into kilometres
e 22,000 mg into grams
f 11,000 g into kg
g 15,000 kg into tonnes
h 36,000,000 g into tonnes
i 7,560,000 ml into litres
j 400,000 cm into km
k 0.023 m into centimetres
l 0.07 l into millilitres
m 4 tonnes into kilograms
n 0.33 kg into grams
o 0.004 tonnes into grams
p 0.001 g into milligrams
q 1.5 km into metres
r 0.0006 m into millimetres

Johann Dase (1824-1861), born in Hamburg, could count objects with the greatest rapidity. With a single glance he could give the number (up to 30 or thereabouts) of peas in a handful scattered on a table; and the ease and speed with which he could count the number of sheep in a herd, of books in a case, or the like, never failed to amaze the beholder. He could give the number of peas in a closed, glass jar by simply shaking the jar and he was the first to find all the prime numbers up to 8 million, which he calculated mentally.

15

NUMBER SPLITTING

This is a very useful device for splitting a difficult sum into two or more easy ones.

ADDITION

 EXAMPLE 1

Suppose you are given the addition sum:

$$
\begin{array}{r}
2\ 3\ 4\ 5 \\
6\ 7\ 3\ 8\ + \\
\end{array}
$$

With 4-figure numbers it looks rather hard.
But if you split the sum into two parts, each part can be done easily and mentally:

$$
\begin{array}{r}
2\ 3\,|\,4\ 5 \\
6\ 7\,|\,3\ 8\ + \\
\hline
9\ 0\,|\,8\ 3 \\
\end{array}
$$

On the right we have $45 + 38$ which (mentally) is 83. So we put this down.
And on the left we have $23 + 67$ which is 90.

EXAMPLE 2

Here is one with decimals, 81.7 + 95.8:

$$
\begin{array}{c}
8\,|\,1\,.\,7 \\
+9\,|\,5\,.\,8 \\
\hline
1\,7\,|\,7\,.\,5
\end{array}
$$

We can split the sum as shown and add 17 + 58, which is 75.
Then 8 + 9 = 17. The point simply goes under the other points.

 EXAMPLE 3

Find 481 + 363.

> You will have to think where to put the line, but it is usually best to put it so
> that there are no carries over the line:

$$
\begin{array}{c}
4\,|\,8\,1 \\
+3\,|\,6\,3 \\
\hline
8\,|\,4\,4 \\
\scriptstyle1
\end{array}
\qquad\qquad
\begin{array}{c}
4\,8\,|\,1 \\
+3\,6\,|\,3 \\
\hline
8\,4\,|\,4
\end{array}
$$

This example is done two in ways. The second way is easier.

 EXAMPLE 4

$$
\begin{array}{c}
6\ 3\ 4\ 3 \\
+\ 2\ 3\ 8\ 3 \\
\hline

\end{array}
$$

It is best to insert two lines here:

$$
\begin{array}{c}
6\,|\,3\ 4\,|\,3 \\
+\ 2\,|\,3\ 8\,|\,3 \\
\hline
8\,|\,7\ 2\,|\,6
\end{array}
$$

 EXERCISE 1

Add the following (try some of them mentally):

a	3 4 5 6	b	1 8 1.9	c	6 4 4 6	d	8 3 2 1	e	7 6 7	f	3 8 3
	4 7 1 7		1 7 1.6		2 8 3 8		1 8 2 3		6 1 6		3 8 4

g	4 4 4	h	8 8 8	i	5 5 1	j	4.5 5 4	k	1 2 3 4	l	5 2 3 4
	2 4 6		7 0 7		6 6 2		3.6 3 6		4 9 4 4		9 3 9 3

m 1884 + 1908   **n** 5555 + 6116   **o** 9119 + 8228   **p** 73839 + 12345

q 2704 + 6889   **r** 843 + 919   **s** 7071 + 7777

SUBTRACTION

 EXAMPLE 5

Consider the subtraction sum:

```
  5 4 5 4
-  1 7 2 6
```

We can split this up into two easy ones:

```
  5 4 | 5 4
-  1 7 | 2 6
  3 7 | 2 8
```

First 54 – 26, which is 28,
then 54 – 17, which is 37.

 EXAMPLE 6

```
  4 4 6 8
-  2 2 8 6
```

Splitting this in the middle like the last one would involve 68 – 86, which is not easy.
So we split as follows:

```
  4 | 4 6 | 8
-  2 | 2 8 | 6
  2 | 1 8 | 2
```

 EXERCISE 2

Subtract the following:

a 3 2 4 3
   1 3 1 9
   _____

b 4 4 4 4
   1 8 2 8
   _____

c 7 0 7 0
   1 5 2 6
   _____

d 3 7 2 1
   1 9 0 9
   _____

e 6 8 8 9
   1 9 3 6
   _____

f 8 5 2
   1 3 9
   _____

g 7 7 7
   5 8 5
   _____

h 6 6 6 6
   2 9 3 8
   _____

i 66.16 – 1.61   **j** 7396 – 1681   **k** 5476 – 2809

MULTIPLICATION

The same technique can be applied here.

 EXAMPLE 7

352 × 2

We can split this sum: 35 / 2 × 2 = 704. 35 and 2 are easy to double.

EXAMPLE 8

Similarly 827×2 becomes $8 / 27 \times 2 = \underline{1654}$,

604×7 becomes $6 / 04 \times 7 = \underline{4228}$,

121745×2 becomes $12 / 17 / 45 \times 2 = \underline{243490}$,

3131×5 becomes $3 / 13 / 1 \times 5 = 15655$.

 EXERCISE 3

Multiply the following:

a 432×3 **b** 453×2 **c** 626×2 **d** 433×3 **e** 308×6

f 814×4 **g** 515×5 **h** 919×3 **i** 1416×4 **j** 2728×2

k 3193×3 **l** 131415×3 **m** 40506×7 **n** 30911×6 **o** 121314×7

DIVISION

Division sums can also often be simplified by this method.

 EXAMPLE 9

The division sum $2\overline{)4\ 3\ 2}$ can be split into: $2\overline{)4} / 32 = \underline{216}$,

because 4 and 32 are both easy to halve.

 EXAMPLE 10

Similarly $2\overline{)3\ 4\ 5\ 6}$ becomes $2\overline{)34} / 56 = \underline{1728}$,

$6\overline{)6\ 1\ 2}$ becomes $6\overline{)6} / 12 = \underline{102}$, (note the 0 here because the 12 takes up two places)

$7\overline{)2\ 8\ 4\ 9}$ becomes $7\overline{)28} / 49 = \underline{407}$,

$3\overline{)2\ 4\ 4\ 5\ 3}$ becomes $3\overline{)24} / 45 / 3 = \underline{8151}$.

 EXERCISE 4

Divide the following mentally:

a $2\overline{)656}$ **b** $2\overline{)726}$ **c** $3\overline{)1821}$ **d** $6\overline{)1266}$ **e** $4\overline{)2048}$ **f** $4\overline{)2816}$

g $3\overline{)2139}$ **h** $2\overline{)2636}$ **i** $2\overline{)56342}$ **j** $3\overline{)91827}$ **k** $2\overline{)387252}$ **l** $4\overline{)812}$

m $6\overline{)4818}$ **n** $8\overline{)40168}$ **o** $3\overline{)15015}$ **p** $13\overline{)39052}$

CHECKING DEVICES

using *The First by the First and the Last by the Last*

It sometimes happens that we only want an approximate value of a calculation and so we do not need to find the exact answer. We may only want to find the first figure of the answer and the appropriate number of noughts.

The First by the First

 EXAMPLE 11

32×41 is approximately <u>1000</u>

by multiplying the first figure of each number together we find that 32×41 is approximately 30×40, which is 1200.
So we expect the answer to be about <u>1000,</u> rounding off to the nearest thousand.

 EXAMPLE 12

Find the approximate value of 641×82.

We want the first figure of the answer and the number of 0's that come after it.
Since $600 \times 80 = 48,000$ and we know the answer will be more than this we can say the answer is about <u>50,000</u> (to the nearest 10,000).

 EXAMPLE 13

Find the approximate value of 39×63.

39 is close to 40 so that *the first by the first* gives $40 \times 60 = 2400$.
And we can say <u>2000</u>.

 EXAMPLE 14

Find an approximate value for 383×88.

$400 \times 90 = 36,000$ and the answer must be below this because both 400 and 90 are above the original numbers, so we can say $383 \times 88 \approx$ <u>30,000</u>.

Note the symbol $\approx$ for **approximately equal to**.

Thus we see that *The First by the First* gives us the first figure of the answer; and the number of figures in the answer is also evident.
We may not always be certain of the first figure (as in the last example) but we will never be more than 1 out.

 EXERCISE 5

Approximate the following:

a 723×81 **b** 67×82 **c** 4133×572 **d** 38×49

e 6109×377 **f** 3333×4444 **g** 1812×1066 **h** 919^2

The Last by the Last

The last figure of a calculation is also apparent by looking at the last figures in the sum.

 EXAMPLE 15

72×83 ends in $\underline{6}$

by multiplying the last figure of each number together we get the last figure of the answer:
$2 \times 3 = 6$.

 EXAMPLE 16

383×887 ends in $\underline{1}$

since $3 \times 7 = 21$, which ends with a 1.

 EXAMPLE 17

$23 \times 48 \times 63$ ends in a $\underline{2}$.

Because 3×8 ends in a 4 and 4×3 ends in a 2.

 EXERCISE 6

What is the last figure of the following sums?

a 456×567 **b** 76543×98 **c** $67 \times 78 \times 89$ **d** $789 + 987$ **e** 715^2 **f** 53^3 **g** 888^4 **h** 23.2^4

CHECKING CALCULATIONS

These two methods of finding an approximate value and the last figure can be used for checking the correctness of an answer.

In the following exercise some of the sums are correct and some are wrong.
Can you find which is which?

 EXERCISE 7

Which of the following sums are correct, judging by the first and last figures?

a $627 \times 762 = 477774$ **b** $715 \times 735 = 525525$ **c** $54 \times 64 = 3456$

d $84 \times 481 = 40404$ **e** $593 \times 935 = 554455$ **f** $592 \times 792 = 468864$

g $726 \times 926 = 672267$ **h** $462 \times 962 = 444442$ **i** $741 \times 777 = 575757$

j $408 \times 842 = 343536$ **k** $733 \times 744 = 545352$ **l** $37 \times 367 = 13579$

m $223 \times 443 = 98789$ **n** $538 \times 539 = 889982$ **o** $265347^2 = 90409030409$

"The knowledge at which geometry aims is knowledge of the eternal."

Plato (429-347 BC)

GREEK PHILOSOPHER

16

POLYGONS AND CO-ORDINATES

✳ You will need a sheet of graph paper.
Draw a horizontal axis along the bottom (this is the long side of your sheet) and number it from 0 to 25 (each unit is 1 cm).
Draw a vertical axis up the left-hand side and number it from 0 to 18.
Put the title POLYGONS at the top.

The **coordinates** of a point is a pair of numbers, in brackets, which describe the position of a point. For example to find the point (3,7) you would go 3 units to the right of the **origin** (this is where your two axes meet) and 7 units up.

 EXERCISE 1

In each of the 7 questions below plot each point and join it to the previous point. Then join the last point to the first to complete the shape. You can use a different colour for each shape if you wish. Also write the name of the shape inside it.

a (1,14) (5,14) (2,17)   triangle (3 sides)

b (7,0) (11,1) (6,4) (5,2)   quadrilateral (4 sides)

c (0,8) (0,12) (7,12) (9,9) (4,5)   pentagon (5 sides)

d (14,12) (17,13) (19,15) (18,17) (14,17) (11,14)   hexagon (6 sides)

e (23,18) (21,16) (21,12) (22,14) (24,14) (25,12) (25,16)   heptagon (7 sides)   ☞

f (25,2) (25,3) (24,4) (21,4) (20,3) (20,2) (21,1) (24,1)   octagon (8 sides)

g (22,9) (20,11) (18,10) (20,9) (18,7) (19,5) (20,7) (22,5) (24,6) (22,7) (24,9) (23,11)

dodecagon (12 sides)

1 Which 2 shapes contain a right-angle?

2 Which 4 shapes have got at least one pair of parallel lines?

3 In the space which is left on your graph paper design and draw your own decagon (10 sides). Can you include a right angle and a pair of parallel lines in the figure?

4 Write down the coordinates of each vertex of your decagon.

5 Make a list in your book of all the polygon shape names you know together with the number of sides it has (remember, a nonagon has 9 sides).

QUADRILATERALS

We have studied triangles in some detail and now we move on to quadrilaterals.
There are many different types of quadrilateral: for example a rectangle has 4 sides so it is a type of quadrilateral.

✴ On a sheet of graph paper number your axes from 0 to 25 along the bottom and 0 to 18 up the left-hand side. Put the title QUADRILATERALS at the top.

EXERCISE 2

Plot each of the following 6 shapes and write the name of the shape inside it.

a (0,0) (3,0) (3,3) (0,3)   square

b (1,13) (1,15) (5,15) (5,13)   rectangle

c (10,0) (12,4) (18.5,4) (16.5,0)   parallelogram

d (25,9) (22,7) (22,4) (25,3)   trapezium

e (25,16) (21,18) (17,16) (21,14)   rhombus

f (4,10) (2,9) (4,4) (6,9)   kite

The descriptions below will help you to decide which type of quadrilateral is which.
✴ Copy them carefully into your book.

Square: has 4 right angles and all sides equal

Rectangle: has 4 right angles

Parallelogram: has opposite sides parallel

Trapezium: has **only 1** pair of parallel sides

Rhombus: has all sides equal

Kite: has a pair of equal sides meeting each other and the other pair also equal (but not equal to the first pair)

So a rectangle is a quadrilateral which has 4 right angles.
But a square is a quadrilateral having 4 right angles, so a square is a rectangle.
This may seem rather strange but it is true. A square is a special kind of rectangle in which all the sides are equal.

✱ Is a square a rhombus?
Yes it is because to be a rhombus a shape has to have all sides equal and this is true for a square.
A square is a special kind of rhombus.

6 Is a parallelogram always a rhombus?

7 Is a rectangle a parallelogram?

8 Is a rectangle a trapezium?

In the remaining space on your graph paper draw 2 more shapes:
9 Draw the kite again, the same size and the same shape, but upside down.

10 Draw the rhombus again, same size and shape, but turned a quarter of a turn so that it points up and down instead of left and right.

✱ Put the name inside the shape.

DIAGONALS OF QUADRILATERALS

A **diagonal** of a figure is a straight line from one vertex (corner) to another vertex which is not next to it.
A diagonal of a quadrilateral simply joins opposite vertices.

So every quadrilateral has 2 diagonals.

Diagonals may   **A** be of **equal** length,

                       B be **perpendicular** (i.e. at right angles to each other),

                       C **bisect** each other (i.e. cut each other exactly in half).

For example, in all rectangles the diagonals must be equal, and they will also bisect each other, but they will not generally be perpendicular.
Look at your rectangle and make sure that you agree with this statement.

  EXERCISE 3

Copy and complete the following table:

	D I A G O N A L S		
	Equal	Perpendicular	Bisect
Square			
Rectangle	yes	no	yes
Parallelogram			
Trapezium			
Rhombus			
Kite			

If possible draw the following:

11 A kite which has 2 right angles.

A trapezium which has 2 right angles.

A trapezium with a pair of opposite sides equal.

You should have 3 drawings.

17

REGULAR POLYGONS AND PERIMETERS

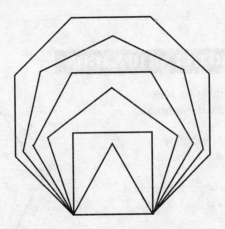

A polygon in which all the angles are equal and all the sides are equal is called a **regular polygon.**

The diagram above shows 6 regular polygons all on the same base line:

        a regular triangle, called an **equilateral triangle,**
        a regular quadrilateral, called a **square,**
        a regular **pentagon,**
        a regular **hexagon,**
        a regular **heptagon** or regular septagon,
        a regular **octagon.**

TO CONSTRUCT AN EQUILATERAL TRIANGLE

You will need a sheet of plain paper, compasses, ruler and pencil.

Draw a horizontal straight line 3cm long leaving at least 3cm of space around it.

Put your compass point on one end of the line
and adjust your compasses so that
the pencil is exactly at the other end.

Draw a circle.

Now put your compass point on the
other end of the line and draw
another circle the same size.

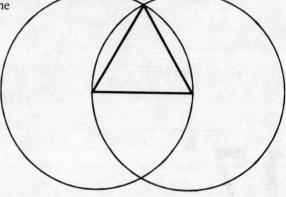

These circles meet at 2 points.
Join the top point to each end of your
straight line and you should have an equilateral triangle.

This is a very simple construction, but it is not necessary to draw complete circles to locate the top point.

1 Draw an equilateral triangle on a base of length 2cm but draw as little as possible of the 2 circles.

TO CONSTRUCT A REGULAR HEXAGON

Start off as before with a straight line 3cm
long but leave about 6cm above it this time.

With your compass set at 3cm, as before,
draw a circle centred at each end of the line.

Put your compass point at the upper intersection
point of these circles and draw a third circle.

This circle meets the first 2 circles at points A and B.
Put your compass point on A and mark a point
on the third circle. Then with the point on B mark
another point on the third circle.

You can now draw in the regular hexagon.

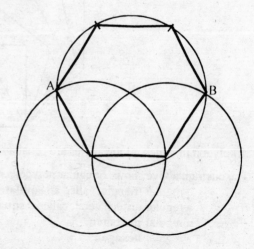

Another way of drawing a regular hexagon
is to draw a circle,
and without moving the compasses,
put the point at an appropriate place on the
circumference (edge) of the circle
and mark a point on the circumference.

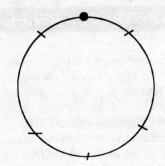

Then put the point on the marked point and
mark another point.

Continue like this until you have gone
right round the circle and then join the 6 points up.

2 Construct a hexagon by this method with your compasses set on 2cm.

TO CONSTRUCT A SQUARE

Draw a base line AB 3cm long leaving about 6cm above it.
Draw a circle, radius 3cm, centred on each end
of the line, as before.

At the upper intersection point C draw a circle
with the same radius.

This circle meets the first 2 circles at D and E.
Put your compass point at D
and then at E and mark the points F, G.

Join A to F and B to G.
Where these 2 lines meet the first 2 circles
indicates the top points of the square.

Draw the square in.

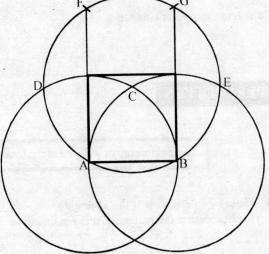

3 Draw another square, this time with a base of 4cm, but do not use heavy or coloured lines
because it will be used in the next section.

TO CONSTRUCT A REGULAR OCTAGON

On your square diagram label the corners A, B, C, D as shown.
We are going to draw a circle round the square
and then find 4 more points on it to form the octagon.

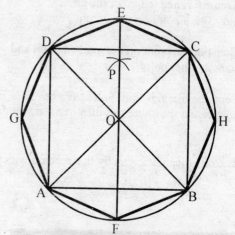

Draw in the diagonals AC and BD.
These meet at O.
Draw a circle centre O, radius OA
so that it passes through A,B,C and D.

To locate P draw an arc, as shown, with
compass point on A and radius AB.

Similarly draw an arc from centre B.

Draw in the vertical line OP (see diagram)
and extend it to locate E and F.

With your compass point on D and radius DE
mark the point G.
Similarly with the point at C mark H.

Complete the regular octagon.

PERIMETERS

The **perimeter** of a figure is the total distance round the outside of it.

✴ What is the perimeter of this rectangle?
   Do not measure the rectangle, it is not drawn to scale.

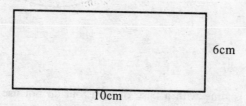

Because the top is the same length as the bottom and the left side is the same as the right side we could find $10 + 6$ and double the result. This makes 32cm.

The word **circumference** means the perimeter of a circle. Circumference is only used in relation to circles.

In this lesson you have constructed:
  A an equilateral triangle of side 3cm,
  B a regular hexagon of side 2cm,
  C a square of side 4cm,
  D a regular octagon.

4 Find the perimeter of each of these figures.
For the octagon you will need to measure a side, but not for the others.

5 What will be the perimeter of a regular pentagon of side 4.5cm?

PERIMETER PROBLEMS

6 Sherlock Holmes Junior has two clues about the sides of a rectangle.
The perimeter of the rectangle is 24cm, and the sides of the rectangle are prime numbers.
Can you deduce the sides of the rectangle?

Knowing that the perimeter is 24 Sherlock Junior deduced that a long and a short side together must be half of this and there is only one pair of prime numbers that add up to 12.

7 In Sherlock's next case the perimeter of the rectangle is 30cm and the sides are known to be in the ratio 4:1 (this means a long side is 4 times longer than a short side).
What are the sides?

You are looking for two numbers that add up to 15, and one of them is 4 times the other.

 EXERCISE 1

Help Sherlock to solve the following cases. Find the:

a perimeter of a rectangle of width 5cm and whose length is twice its width,

b sides of a rectangle if the perimeter, $P = 24$cm and the length is 3 times the width,

c sides of a rectangle if $P = 26$cm and the length is 1cm more than the width,

d sides of a triangle where P = 20cm, the base is 8cm and the other 2 sides are equal,

e sides of a square whose perimeter is twice that of a square of perimeter 24cm,

f sides of a quadrilateral whose sides are consecutive whole numbers, and P = 30,

g perimeter of a square whose sides are half those of a square of perimeter 12,

h sides of a square whose perimeter is twice that of a rectangle of sides 6 and 8.

i A triangle has a base 13cm long, the other sides being 9 and 10cm.

    Find **i** its perimeter,
            ii the sides of a square with equal perimeter,
           iii the sides of a rectangle of equal perimeter, given that the length is 3 times the width,
           iv the perimeter of a square constructed on the base of the triangle.

j Find the perimeter of a rectangle which has twice the height and half the base of a square of side 6cm.

k Find the perimeter of the largest square that will fit into a rectangle 7cm by 12cm.

l Three of the sides of a pentagon with sides 3, 5, 8, 9, 12cm are used to form the sides of a triangle of perimeter 23cm. Which ones are used?

18

ALL FROM 9 AND THE LAST FROM 10

ALL FROM 9

Suppose we apply this part of the Sutra to a few numbers.

 EXAMPLE 1

Apply *All from 9* to the number 876.

We take each of the figures in this number from 9:

8 from 9 is **1**,   7 from 9 is **2**,   6 from 9 is **3**,

so we get the number **123**.

Similarly   4 9 7   7 8 6 6   6 4   2
                 5 0 2   **2 1 3 3**   **3 5**   **7**

✳ Check that you agree with the examples above.
✳ Notice that the columns all add up to 9.

 EXERCISE 1

Apply *All from 9* to the following numbers:

a 356   **b** 86413   **c** 1918   **d** 70891   **e** 60

ALL FROM 9 AND THE LAST FROM 10

> This is the same except that the very last figure is taken from 10 instead of 9.

 EXAMPLE 2

If we apply the full Sutra to 876 we get **124** this time, because the 6 at the end is take from 10.

Similarly	3883	64	98	6	10905
	6117	**36**	**02**	**4**	**89095**

 EXAMPLE 3

Applying the Sutra to 3450 or any number that ends in 0 we need to be a bit careful. If we take the last figure as 0 here, when we take it from 10 we get 10 which is a 2-figure number. To avoid this we **take 5 as the last figure**: we apply the Sutra to 345 and simply put the 0 on afterwards. So we get **6550**.

EXERCISE 2

Apply *All from 9 and the Last from 10* to the following:

a 444   **b** 975   **c** 2468   **d** 18276   **e** 8998   **f** 10042

g 1020304   **h** 9888   **i** 52805   **j** 6770   **k** 9631   **l** 35700

m 1234560 **n** 7   **o** 3300   **p** 40   **q** 314159   **r** 1080

Look again at the number in part **a** above.
Your answer should be 556.
Now add the number and its '*All from 9...*' number:

$$
\begin{array}{r}
4\ 4\ 4 \\
5\ 5\ 6 + \\
\hline
1\ 0\ 0\ 0 \\
\end{array}
$$ you get 1000.

❋ Try this with at least three other numbers from Exercise 2.
1 What does the number and its '*All from 9...*' number add up to?

You should find that the total is always 10, 100, 1000, 10000 . . .
The total is always one of these unities, called **base numbers**.

This means that when we applied the Sutra to 444 we found how much it was below 1000.
Or, in other words, the Sutra gave the answer to the sum 1000 – 444. The answer is 556.

> The formula *All from 9 and the Last from 10*
> subtracts numbers from the next highest unity

 EXAMPLE 4

1000 – 864 = 136 Just apply *All From 9 and the Last from 10* to 864,

1000 – 307 = 693,

10000 – 6523 = 3477,

100 – 76 = 24,

1000 – 580 = 420,   remember: apply the Sutra just to 58 here.

In every case here the number is being subtracted from its next highest unity.

 EXERCISE 3

 . . . 9   9   9   9   10

Subtract the following:

a 1000 – 481   **b** 1000 – 309   **c** 1000 – 891   **d** 1000 – 976

e 100 – 78   **f** 100 – 33   **g** 100 – 61   **h** 10000 – 8877

i 10000 – 1357   **j** 10000 – 9876   **k** 1000 – 808   **l** 1000 – 710

m 1000 – 930   **n** 10000 – 6300   **o** 1000 – 992   **p** 100000 – 65856

In all of the above sums you may have noticed that the number of zeros in the first number is the same as the number of figures in the number being subtracted.
e.g. 1000 – 481 has three zeros and 481 has three figures.

FIRST EXTENSION

 EXAMPLE 5

Suppose we had   1000 – 43.
This has three zeros, but 43 is only a 2-figure number.
We can solve this by writing   1000 – 043 = <u>957</u>.
We put the extra zero in front of 43, and then apply the Sutra to 043.

 EXAMPLE 6

10000 – 58.
Here we need to add two zeros: 10000 – 0058 = <u>9942</u>.

In the following exercise you will need to insert zeros, but you can do that mentally if you like.

 EXERCISE 4

Subtract the following:

a 1000 – 86   **b** 1000 – 29   **c** 1000 – 77   **d** 1000 – 93   **e** 1000 – 35

f 10000 – 678   **g** 10000 – 353   **h** 10000 – 177   **i** 10000 – 762   **j** 10000 – 62

k 10000 – 85   **l** 10000 – 49   **m** 1000 – 8   **n** 10000 – 3

SECOND EXTENSION

 EXAMPLE 7

Now consider   600 – 77.

We have 600 instead of 100.
In fact the 77 will come off one of those six hundreds, so that 500 will be left.
So   600 – 77 = <u>523</u>.
The 6 is reduced by one to 5, and the Sutra is applied to 77 to give 23.

This reduction of one illustrates the Sutra *By One Less than the One Before*.

 EXAMPLE 8

5000 – 123 = <u>4877</u>   the 5 is reduced by one to 4,
                    and the Sutra converts 123 to 877.

 EXAMPLE 9

Subtractions like this frequently arise when buying things.
If we offer a £20 note when paying
for goods costing £3.46 the change we get should be <u>£16.54</u>.

We could think of the sum as being the same as 2000 – 346 (in pence), so the 2 is reduced to 1, and *All from 9...* is applied to 346.

 EXAMPLE 10

8000 – 3222

Considering the thousands there will be 4 left in the thousands column
because we are taking over 3 thousand away.

All from 9... is then applied to the 222 to give 778.

So 8000 – 3222 = <u>4778</u>.

EXERCISE 5

Subtract the following:

a 600 – 88   **b** 400 – 83   **c** 900 – 73   **d** 500 – 31

e 700 – 46   **f** 200 – 69   **g** 3000 – 831   **h** 6000 – 762

i 8000 – 3504   **j** 5000 – 1234   **k** 300 – 132   **l** 2000 – 1444

m 2000 – 979   **n** 50000 – 4334   **o** 70000 – 8012   **p** £20 – £7.44

q £20 – £12.78   **r** £10 – £3.18   **s** £50 – £28.05

COMBINING THE FIRST AND SECOND EXTENSIONS

 EXAMPLE 11

6000 – 32

You will see here that we have a 2-figure number to subtract from 6000 which has three zeros. We will need to apply both of our extensions together.
The sum can be written   6000 – 032.
Then   6000 – 032 = 5968.
The 6 is reduced to 5, and the Sutra converts 032 to 968.

 EXAMPLE 12

30000 – 63 = 30000 – 0063 = 29937   the 3 becomes 2, and 0063 becomes 9937.

> try some mentally

 EXERCISE 6

Subtract the following:

a 5000 – 74   **b** 8000 – 58   **c** 3000 – 43   **d** 7000 – 81

e 6000 – 94   **f** 4000 – 19   **g** 80000 – 345   **h** 30000 – 276

i 50000 – 44   **j** 700 – 8   **k** 800 – 6   **l** 400 – 3

m 30000 – 54   **n** 20000 – 222   **o** 30000 – 670   **p** 70000 – 99

The final exercise is a mixture of all the types we have met:

 EXERCISE 7

Subtract:

a 100 – 34	**b** 1000 – 474	**c** 5000 – 542	**d** 800 – 72
e 1000 – 33	**f** 5000 – 84	**g** 700 – 58	**h** 9000 – 186
i 10000 – 4321	**j** 200 – 94	**k** 10000 – 358	**l** 400 – 81
m 7000 – 88	**n** 900 – 17	**o** 30000 – 63	**p** 90000 – 899

PASCAL'S TRIANGLE

The French mathematician Blaise Pascal wrote a book about a triangle of numbers, part of which is shown below.

```
              1
            1   1
          1   2   1
        1   3   3   1
      1   4   6   4   1
    1   5   10  10  5   1
```

You will see various patterns if you study this.

There is a vertical line of symmetry and every row has one more number than the one before.
Each number is formed from the line above by adding the two numbers just above it: 10 is 4+6 for example. This means the triangle can be extended as far as you like.

✳ Copy this array of numbers and extend it by adding 3 more rows.

If you look at the diagonal rows of numbers you will see the rows of 1's. And the next diagonal is also a familiar sequence: 1, 2, 3, 4. . . .

1 Starting at the top of this 1,2,3,4... diagonal write down the sum of
  a) the first 2 numbers,
  b) the first 3 numbers,
  c) the first 4 numbers,
  d) the first 5 numbers.

2 What is the name for this sequence of numbers?

3 Where can you see this sequence in Pascal's Triangle?

Now look at the next diagonal of numbers: 1, 3, 6, 10 . . .

4 Write down the sum of the first 2, 3, 4 and 5 numbers.
  What do you notice about your answers?

✳ Starting with any **1** on the outside of Pascal's Triangle
✳ Add up some consecutive numbers in the diagonal and look for the total in the Triangle.
  For example:
  1+3+6+10 (shown opposite) = 20.
  And 20 is found below 10 in the next row.

✳ Take another diagonal, starting with 1, and add up the first 1, 2, 3 etc. numbers in that diagonal and you will always find the total just below the end of the diagonal.

5 Write down two examples of this.

```
            1
          1   1
        1   2   1
      1   3   3   1
    1   4   6   4   1
  1   5   10  10  5   1
1   6   15  20  15  6   1
```

The Fibonacci Sequence is a famous sequence of numbers formed by starting with the numbers **1** and **1** and forming new numbers by always adding the last two numbers in the sequence to get the next number.

$$1 \quad 1 \quad 2 \quad 3 \quad 5 \quad 8 \ldots$$

So 1+1 = 2 (the 3rd number), 1 + 2 = 3 (the 4th number), 2+3 = 5 (the 5th number) etc..

6 Copy the sequence and its name and extend it until you have 12 numbers.

The numbers in this sequence arise frequently in nature, in the petals of flowers, the spirals of cones and sunflowers, musical scales etc..
It also has many remarkable mathematical properties.

7 For example, take any three consecutive numbers in the sequence. Write them down.
   Multiply the outer numbers together and write down the answer.
   Square the middle number and write down the answer.
   Choose three other consecutive numbers and do the same.
   What do you notice?
   Check your finding by looking at some more sets of
   three consecutive numbers.

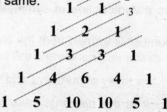

In the Triangle shown opposite other diagonals
are shown by the dashed lines.
The first three diagonals total 1, 2 and 3, as shown.

$$
\begin{array}{ccccccc}
 & & & 1 & & & \\
 & & 1 & & 1 & & \\
 & 1 & & 2 & & 1 & \\
1 & & 3 & & 3 & & 1 \\
\end{array}
$$

1   4   6   4   1

1   5   10   10   5   1

8 Find the sums for the next four diagonals.

   What do you find?

Here is a variation on Pascal's Triangle.
The outside diagonals are **1** on the left and **2** on the right.
But as before the rule for obtaining the other numbers
is to add the two numbers on the row above.

$$
\begin{array}{cccccc}
 & & & & 1 & 2 \\
 & & & 1 & 3 & 2 \\
 & & 1 & 4 & 5 & 2 \\
 & 1 & 5 & 9 & 7 & 2 \\
1 & 6 & 14 & 16 & 9 & 2 \\
\end{array}
$$

9 Copy the triangle, check this rule is obeyed and extend the triangle for three more rows.

✳ Add the dashed lines on your triangle, like the ones shown in the triangle above this one, and add up the diagonals in the same way. The first three totals should be 1, 3, 4.

10 Obtain the first 8 numbers of this sequence.
   You should find that it is like the Fibonacci sequence: adding two consecutive numbers gives the next number in this sequence.

11 Use this rule to extend the sequence for four more terms.

12 Next to the diagonal of 2's in the above triangle there is another diagonal.
   You should recognise this sequence. What are these numbers called?

13 Next to this diagonal is another sequence which you should also recognise.
   What are these numbers called?

"*Vedic Mathematics is the balancing power between two opposite qualities of its own nature— unifying and diversifying.*"

Maharishi Mahesh Yogi

FOUNDER OF TRANSCENDENTAL MEDITATION

19

BAR NUMBERS

The number 19 is very close to 20.

And it can therefore be conveniently written in a different way: as $2\bar{1}$

$2\bar{1}$ means $20 - 1$, the minus is put on top of the 1.

Similarly $3\bar{1}$ means $30 - 1$ or 29.

And $4\bar{2}$ means 38.

This is like telling the time when we say 'ten to seven' or 'five to seven' instead of 6:50 or 6:55.

We pronounce $4\bar{2}$ as "four, bar two" because the 2 has a bar on top.

➡ EXAMPLE 1

$7\bar{2} = 68$,

$86\bar{1} = 859$, because $6\bar{1} = 59$ (the 8 is unchanged),

$127\bar{2} = 1268$, because $7\bar{2} = 68$,

$6\bar{3}0 = 570$, because we have $600 - 30$.

 EXERCISE 1

Convert the following numbers:

a $6\bar{1}$ **b** $8\bar{2}$ **c** $3\bar{3}$ **d** $9\bar{1}$ **e** $7\bar{4}$ **f** $9\bar{5}$ **g** $5\bar{7}$ **h** $46\bar{2}$ ☞

i 85$\bar{1}$ **j** 77,4$\bar{1}$ **k** 999$\bar{1}$ **l** 1$\bar{2}$ **m** 11$\bar{1}$ **n** 12$\bar{3}$ **o** 3$\bar{4}$0

We may also need to put numbers **into** bar form.

 EXAMPLE 2

79 = 8$\bar{1}$ because 79 is 1 less than 80,

239 = 24$\bar{1}$ because 39 = 4$\bar{1}$,

7689 = 769$\bar{1}$ because 89 = 9$\bar{1}$.

508 = 51$\bar{2}$ 08 becomes 1$\bar{2}$

? **EXERCISE 2**

Put the following into bar form:

a 49   **b** 58   **c** 77   **d** 88   **e** 69   **f** 36   **g** 17

h 359   **i** 848   **j** 7719   **k** 328   **l** 33339   **m** 609   **n** 708

EXAMPLE 3

How would you remove the bar number in 5$\bar{1}$3 ?
The best way is to split the number into two parts: 5$\bar{1}$/3.
Since 5$\bar{1}$ = 49, the answer is **493**

> If a number has a bar number in it split the number after the bar.

EXAMPLE 4

7$\bar{3}$1 = 7$\bar{3}$/1 = **671**,

52$\bar{4}$2 = 52$\bar{4}$/2 = **5162**,

3$\bar{2}$15 = 3$\bar{2}$/15 = **2815**   since 3$\bar{2}$ = 28,

5$\bar{1}$3$\bar{2}$ = 5$\bar{1}$/3$\bar{2}$ = **4928**   since 5$\bar{1}$ = 49 and 3$\bar{2}$ = 28,

3$\bar{1}$3$\bar{2}$3$\bar{3}$ = 3$\bar{1}$/3$\bar{2}$/3$\bar{3}$ = **292827**.

❓ EXERCISE 3

Remove the bar numbers:

a $6\bar{1}4$ b $4\bar{2}3$ c $5\bar{2}5$ d $3\bar{1}7$ e $45\bar{2}3$ f $23\overline{42}5$ g $2\bar{2}2$

h $333\bar{2}3$ i $5\bar{1}32$ j $2\bar{3}55$ k $5\bar{4}4321$ l $4\bar{1}3\bar{1}$ m $6\bar{2}7\bar{3}$ n $2\bar{1}1$

o $41\bar{3}1$ p $52\bar{3}3$ q $7\bar{1}52$ r $1\bar{3}15\bar{1}$ s $9\bar{2}83$ t $1\bar{3}1$ u $13\bar{1}51$

☺ BAR NUMBERS GAME   For 2 to 4 players.

Put all the red cards on the table in 3 rows of 6, face up.
Put the white pack in the middle with the numbers facing down.
The first player takes a card from the top of the pack and looks for the equivalent card on the table.
If the player picks the correct card both cards are kept.
If not correct the player puts the white card at the bottom of the pack.
If there is a dispute about whether the chosen card is right or not the answers may be consulted.
The next player then takes a white card and so on.
When all the cards are gone the game is repeated with the white cards laid out and the red ones in a pack.
When all these cards are gone the player with the most cards wins.

ALL FROM 9 AND THE LAST FROM 10

So far we have only had a bar on a single figure.
But we could have two or more bar numbers together.

 EXAMPLE 5

Remove the bar numbers in $5\overline{33}$.

The 5 means 500, and $\overline{33}$ means 33 is to be subtracted.

So $5\overline{33}$ means 500 – 33, and we have met sums like this in the last chapter.

$500 - 33 = \textbf{467}$ because the 33 comes off one of the hundreds, so the 5 is reduced to 4.

And applying **All from 9 and the Last from 10** to 33 gives 67.

EXAMPLE 6

Similarly $7\overline{14} = \mathbf{686}$ the 7 reduces to 6 and the Sutra converts 14 to 86,

$26\overline{21} = \mathbf{2579}$ 26 reduces to 25,

$7\overline{02} = \mathbf{698}$ the Sutra converts 02 to 98,

$50\overline{3} = \mathbf{497}$ 50 is reduced to 49 (alternatively, write $50\overline{3}$ as $5\overline{03}$ - see previous example),

$4\overline{2}0 = 4\overline{2}\,0 = \mathbf{380}$

EXAMPLE 7

$4\overline{23}1 = \mathbf{3771}$

Here we can split the number after the bar: $4\overline{23}/1$.
$4\overline{23}$ changes to 377, and we just put the 1 on the end: $4\overline{23}1 = \mathbf{3771}$

EXAMPLE 8

Similarly $5\overline{12}4 = 5\overline{12}/4 = \mathbf{4884}$,

$3\overline{11}33 = 3\overline{11}/33 = \mathbf{28933}$,

$5\overline{123} = \mathbf{4877}$,

$3\overline{1}4\overline{31} = 3\overline{1}/4\overline{31} = \mathbf{29369}$.

? EXERCISE 4

Remove the bar numbers:

a $6\overline{12}$	**b** $7\overline{33}$	**c** $2\overline{31}$	**d** $5\overline{11}$	**e** $9\overline{04}$	**f** $70\overline{6}$
g $55\overline{23}$	**h** $72\overline{41}$	**i** $333\overline{22}$	**j** $6\overline{21}4$	**k** $5\overline{31}22$	**l** $33\overline{22}44$
m $7\overline{333}$	**n** $6\overline{123}$	**o** $5\overline{104}$	**p** $44\overline{112}$	**q** $74\overline{031}$	**r** $7\overline{103}1$
s $6\overline{3322}$	**t** $3\overline{1102}$	**u** $4\overline{222}3$	**v** $3\overline{1141}$	**w** $3\overline{2122}$	**x** $310\overline{23}$
y $37\overline{20}$	**z** $400\overline{3}$				

SUBTRACTION

These bar numbers give us an alternative way of subtracting numbers.

 EXAMPLE 9

$$\begin{array}{r} 4\,4\,4 \\ -\,2\,8\,6 \end{array}$$

Subtracting in each column we get $4-2=2$, $4-8=-4$, $4-6=-2$.
Since these negative answers can be written with a bar on top we can write:

$$\begin{array}{r} 4\,4\,4 \\ -\,2\,8\,6 \\ \hline 2\,\bar{4}\,\bar{2} \end{array}$$

and $2\,\bar{4}\,2$ is easily converted into **158**

 EXAMPLE 10

Similarly
$$\begin{array}{r} 6767 \\ -\ 1908 \\ \hline 5\bar{2}6\bar{1} \end{array} = \textbf{4859}$$

❓ EXERCISE 5

Subtract using bar numbers:

a 5 4 3
  1 6 8 –
  ———

b 5 6 7
  2 7 9 –
  ———

c 8 0 4
  3 8 8 –
  ———

d 7 3 7
  5 5 8 –
  ———

e 6 4 1 3
  1 8 7 8 –
  ———

f 8 0 2 4
  5 3 3 9 –
  ———

g 6 5 4 3
  2 8 8 1 –
  ———

h 7 1 0 3
  3 9 9 1 –
  ———

i 4 5 4 5
  1 7 9 1 –
  ———

j 3 2 0 4
  2 0 8 1 –
  ———

k 5 6 4 2 3
  2 8 1 7 2 –
  ———

l 3 4 5 6 7
  3 0 9 0 9 –
  ———

m 4 5 6 5 4
  2 7 9 8 6 –
  ———

n 6 3 3
  8 8 –
  ———

o 8 2 2
  5 7 7 –
  ———

p 5 5 5
  2 7 5 –
  ———

q 4 5 5 0
  1 2 3 4 –
  ———

r 8 0 2 4
  5 3 2 9 –
  ———

CREATING BAR NUMBERS

One of the main advantages of bar numbers is that we can remove high digits in a number.
For example writing **19** as **2$\bar{1}$** means we do not have to deal with the large 9.

EXAMPLE 11

Remove the large digits from 287.

Here the 8 and the 7 are large (we say that 6, 7, 8, 9 are large digits).

So we write 287 as $3\overline{1}\overline{3}$ the 2 at the beginning is increased to 3, and the Sutra is applied to 87 to give 13.

You will agree that 287 is 13 below 300, which is what $3\overline{1}\overline{3}$ says.

 EXAMPLE 12

Similarly $479 = 5\overline{2}\overline{1}$,

$3888 = 4\overline{1}\overline{1}\overline{2}$,

$292 = 3\overline{1}2$,

$4884 = 5\overline{1}\overline{2}4$,

$77 = 1\overline{2}\overline{3}$ (you can think of 77 as 077),

and so on.

 EXERCISE 6

Remove the large digits from the following:

a 38	**b** 388	**c** 298	**d** 378	**e** 4887	**f** 39876
g 3883	**h** 1782	**i** 3991	**j** 3822	**k** 4944	**l** 390
m 299	**n** 98	**o** 87	**p** 888	**q** 996	**r** 2939
s 1849	**t** 7	**u** 3827	**v** 191919	**w** 1996	

"The advancement and perfection of Mathematics are intimately connected with the prosperity of the state."

Napoleon I (1769-1821)

20

ON THE FLAG

CALCULATING FROM LEFT TO RIGHT

It is common to do calculations starting at the right and working towards the left.
This is however not always the best way.
Calculating from left to right is much easier, quicker and more useful.

The reason for this is that numbers are written and spoken from left to right.
Also in calculations we often only want the first one, two or three figures of an answer, and starting on the right we would have to do the whole sum and so do a lot of useless work.

In this chapter all the calculations will be done mentally.
We will write down only the answer.

ADDITION

 EXAMPLE 1

Given the addition sum

$$\begin{array}{r} 2\ 3 \\ 4\ 5\ + \\ \hline \end{array}$$

there is no difficulty in finding the answer.
From left to right the columns add up to $\underline{6}$ and $\underline{8}$.
So the answer is $\underline{68}$.

EXAMPLE 2

But in the sum   4 5
                                                  3 8 +
                                                  ____

the totals we get are **7** and **13**, and 13 is a 2-figure number.
The answer is not **713**: the 1 in the 13 must be carried over and added to the **7**.
This gives <u>83</u> as the answer.

This is easy enough to do mentally, we add the first column and increase this by 1 if there is a carry coming over from the second column. Then we tag the last figure of the second column onto this.

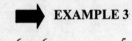 **EXAMPLE 3**

    6   6   5   5   8   4   5   6
    2   8 +   3   5 +   5   8 +   9   6 +
    9   4   9   0   1 4   2   1 5   2

    8,1 4 = 94   8,1 0 = 90   1 3,1 2 = 142   1 4,1 2 = 152

> In every case the tens figure in the right-hand column total
> is carried over to the left-hand column total.

We use the curved lines to show which figures are to be combined.

 EXERCISE 1

Add the following mentally from left to right:

a 5 6   **b** 8 8   **c** 4 5   **d** 5 4
       6 7 +   3 3 +   6 7 +   6 4 +
       ____   ____   ____   ____

e 3 9   **f** 2 7   **g** 7 7   **h** 6 3
       4 9 +   5 6 +   8 8 +   7 4 +
       ____   ____   ____   ____

 EXAMPLE 4

                                          1 8 7
                                          4 4 6 +
                                          _____
                                                              ☞

Here the three column totals are **5, 12** and **13** so two carries are needed.

The 1 in the 12 will be carried over to the 5 making it a 6.
So when the 5 and the 12 are combined we get **62**.
The 1 in the 13 is then carried over and added onto the 2 in 62, making it 63.
So combining 62 and 13 gives the answer, **633**.

It is important to get the idea of doing this mentally from left to right:

First we think of 5, the first total.

Then we have 5,12 **which we mentally combine into 62.**

Hold this 62 in the mind, and with the third total we have 62,13

which becomes 633.

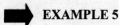 **EXAMPLE 5**

$$
\begin{array}{r}
7\ 7\ 7 \\
4\ 5\ 6\ + \\
\hline
\end{array}
$$

The first two columns give 11,12 which becomes 122.

Then with the third column we have 122,13 which is **1233**.

 **EXAMPLE 6**

$$
\begin{array}{r}
5\ 5\ 5\ 5 \\
3\ 1\ 3 \\
6\ 2\ 4\ + \\
\hline
\end{array}
$$

Starting at the left we have 5,14 = 64.

Then 64,8 = 648 (**there is no carry here** as 8 is a single figure).

Finally 648,12 = **6492**.

 EXERCISE 2

Add the following sums mentally from left to right:

a 3 6 3
   4 5 6 +
   ─────

b 8 1 9
   9 1 8 +
   ─────

c 7 7 7
   4 4 4 +
   ─────

d 7 3 7
   1 3 9 +
   ─────

e 3 4 5
   9 3 7 +
   ─────

f 1 3 6 9
   3 8 8 3 +
   ───────

g 9 6 3 1
   8 7 0 9 +
   ───────

h 4 4 4 4
   4 8 3 8
     5 5 5 +
   ───────

On the flag

EXERCISE 3

In this exercise you should work with a friend.
One of you will give your friend five sums which you will speak out loud (for example "add sixty six and seventy seven"). Your friend will add these mentally from left to right and tell you the answer. You should decide if it is right and correct any mistake.
After five sums you will change over and your friend will give you five sums.
The sums are chosen to be easy to remember.

a $55 + 77$ **b** $86 + 86$ **c** $39 + 38$ **d** $48 + 84$ **e** $71 + 82$

f $66 + 88$ **g** $78 + 78$ **h** $47 + 48$ **i** $57 + 75$ **j** $63 + 74$

If you are confident about this and you are getting them right you can try the harder sums below. Five each again. Repeat the sum if necessary.

 EXERCISE 4

Mentally add:

a $66 + 78$ **b** $92 + 79$ **c** $555 + 888$ **d** $616 + 717$ **e** $789 + 789$

f $77 + 69$ **g** $83 + 68$ **h** $777 + 555$ **i** $828 + 626$ **j** $678 + 678$

In all these sums the numbers are held in the mind (*On the Flag*) and built up digit by digit until the answer is complete.

MULTIPLICATION

➡️ **EXAMPLE 7**

Suppose we have the sum:

$$\begin{array}{r} 2\ 3\ 7 \\ 2 \times \\ \hline \\ \hline \end{array}$$

We multiply each of the figures in 237 by 2 starting at the left.
The answers we get are **4, 6, 14**.

Since the 14 has two figures the 1 must be carried leftwards to the 6.
So $4, 6, 14 = \underline{\textbf{474}}$.

Again we build up the answer mentally from the left: first 4, then 4,6=46, then $4, 6, 14 = 474$.

EXAMPLE 8

$$\begin{array}{r} 2\ 3\ 6 \\ 7\ \times \\ \hline \\ \hline \end{array}$$

First we have 14,
then 14,21 = 161,
then 161,42 = **1652**

 EXAMPLE 9

For 73 × 7 we get 49,21 = **511**   (because 49+2 = 51)

 EXERCISE 5

Multiply the following from left to right:

a 2 7	**b** 7 6	**c** 2 6	**d** 7 2	**e** 7 8	**f** 8 3
3 ×	6 ×	6 ×	7 ×	9 ×	3 ×

g 6 4 2	**h** 2 5 6	**i** 7 4 1	**j** 2 2 3
4 ×	3 ×	3 ×	9 ×

k 1 0 5 9	**l** 8 6 3 1	**m** 5 4 3 2	**n** 4 0 9 7
7 ×	4 ×	8 ×	7 ×

You should now practice these without seeing the sum.
Give your friend the first five sums in the following exercise, then your friend can give five to you.

 EXERCISE 6

Multiply mentally from left to right:

a 33 × 4	**b** 55 × 7	**c** 34 × 8	**d** 66 × 6	**e** 62 × 4
f 55 × 5	**g** 44 × 4	**h** 43 × 7	**i** 88 × 4	**j** 73 × 3

And to finish, if you found these fairly straight forward, here are some harder sums.

 EXERCISE 7

a 78 × 6	**b** 77 × 7	**c** 444 × 4	**d** 345 × 7	**e** 373 × 7
f 69 × 7	**g** 79 × 9	**h** 666 × 4	**i** 246 × 8	**j** 292 × 9

21

PRIME AND COMPOSITE NUMBERS

We are familiar with prime numbers like 7, 11, 13 that have no factors except 1 and themselves.

> Numbers that are not prime are called **composite numbers** because they can be expressed as **a product of prime factors**.

 EXAMPLE 1

Express 6 as a product of prime numbers.

The number 6 is not a prime number. It is therefore a composite number.
We need to write 6 as a product of prime numbers.
We can say $6 = 1 \times 6$ or $6 = 2 \times 3$, but since 2 and 3 are prime numbers and 6 is not
$6 = 2 \times 3$ is the answer.

 EXERCISE 1

Write the following numbers as a product of primes:

a 14   **b** 33   **c** 21   **d** 35   **e** 9

Expressing a number as a product of prime numbers shows up the inner structure of the number and this can be very useful.

FACTOR TREES

 EXAMPLE 2

Write 12 as a product of prime numbers.

You will not find two prime numbers whose product is 12.

The answer is in fact 12 = 2×2×3.

✳ Check that you agree that 2×2×3 = 12.

The answer is not so obvious in this case.
One way of getting the answer is to use a **Factor Tree**:

Think of two numbers whose product is 12, say 6×2, and draw a sketch:

$$12 \begin{array}{c} 6 \\ 2 \end{array}$$

Now of 6 and 2, 6 is not prime.
But 6 = 2×3, so we extend the tree:

Now the numbers at the tips of the branches are all prime, so we get 12 = 2×2×3.

The order of the numbers in the answer does not matter as we get 12 whatever order we have, but it is usual to put the numbers in ascending order.

 EXAMPLE 3

Write 60 as a product of prime factors.

Suppose we start with 60 = 4×15:

Neither 4 nor 15 is prime
so we put extra branches on both:

Now we see that all the end points are prime.

So we can write 60 = 2×2×3×5.

✳ Check that 2×2×3×5 gives 60.
✳ Starting with another product (instead of 4×15) which makes 60, construct a factor tree.
   You should still get the same answer as in the example above.

 EXERCISE 2

Express the following as a product of prime numbers:

a 36   **b** 28   **c** 100   **d** 48   **e** 84   **f** 32

AN ALTERNATIVE METHOD

The factor tree method can lead to rather large trees if the number is very large.
The following method avoids the need for drawing the tree.

 EXAMPLE 4

Write 200 as a product of prime factors.

We divide 200 by the first prime number, 2, until it cannot be divided any more by 2, the divide in the same way by the next prime number, and so on.

So we write $200 = 2×100$,
              $= 2×2×50$ as $100 = 2×50$,
              $= 2×2×2×25$ as $50 = 2×25$,
              $= 2×2×2×5×5$ as $25 = 5×5$.

So $\underline{200 = 2×2×2×5×5}$ which can also be written as $200 = 2^3×5^2$.

 EXERCISE 3

Express as a product of prime factors (use any method you like):

a 4   **b** 9   **c** 8   **d** 14   **e** 16   **f** 18

g 24   **h** 39   **i** 42   **j** 54   **k** 64   **l** 66

m 72   **n** 108

In fact there is an important **theorem** (law) in arithmetic which states that every whole number is prime or else can be written as a product of prime numbers in only one way.

So, for example, since $12 = 2×2×3$ there is no different product of prime numbers that make 12: it is the only answer.

HIGHEST COMMON FACTOR: HCF

Suppose we have two numbers: 70 and 99.

Expressing each as a product of prime factors we find: $70 = 2 \times 5 \times 7$, $99 = 3 \times 3 \times 11$.

Examining these factors we see that there is no common factor (except the number 1, which is a factor of every whole number). That is to say there is no factor of one number which is also a factor of the other number, except for 1.

Such a pair of numbers are said to be **Relatively Prime**, they are prime in relation to each other.

Similarly the three numbers 6, 10, 15 are relatively prime because there is no factor which divides into all three of the numbers.

✳ Check that you agree with this.

Now suppose we have the numbers 18 and 30.

Breaking them down into prime factors: $18 = 2 \times 3 \times 3$, $30 = 2 \times 3 \times 5$.

So 18 and 30 are not relatively prime, they have factors in common.

This means that both numbers can be divided by **2**, by **3** and by $2 \times 3 = $ **6**.

Of these three factors the number 6 is the **Highest Common Factor** or HCF.

> We find the Highest common factor of two or more numbers by
> expressing each number as a product of prime factors,
> and multiplying together all the factors which the prime factors have in common.

 EXAMPLE 5

Find the highest common factor of 30 and 75.

First find the prime factors: $30 = 2 \times 3 \times 5$, and $75 = 3 \times 5 \times 5$.
3×5 is common to both so the highest common factor is <u>15</u>.
15 is the largest number that divides into both 30 and 75.

 EXAMPLE 6

Find the highest common factor of 48 and 72.

$48 = 2^4 \times 3$, $72 = 2^3 \times 3^2$ so the HCF is $2^3 \times 3$ which is <u>24</u>.

EXAMPLE 7

Find the HCF of 24 and 75

$24 = 2^3 \times 3$, $75 = 3 \times 5^2$ therefore $\underline{HCF = 3}$.

 EXAMPLE 8

Find the HCF for 140 and 27

$140 = 2 \times 2 \times 5 \times 7$, $27 = 3 \times 3 \times 3$ therefore $\underline{HCF = 1}$ and the numbers are relatively prime.

EXAMPLE 9

Find the HCF for 42, 63, 90

$42 = 2 \times 3 \times 7$, $63 = 3 \times 3 \times 7$, $90 = 2 \times 3 \times 3 \times 5$ so $\underline{HCF = 3}$.

EXERCISE 4

Find the highest common factor for the following numbers (you may find you can write some of the answers down after just looking at the numbers):

a 36, 56	b 42, 60	c 28, 63	d 30, 96
e 48, 64	f 48, 84	g 80, 63	h 50, 120
i 19, 72	j 29, 59	k 44, 46	l 21, 63
m 98, 84	n 143, 52	o 240, 168	p 315, 270
q 18, 24, 60	r 36, 42, 80	s 30, 35, 54	t 108, 224, 225

BY ADDITION AND BY SUBTRACTION

Another useful result, when numbers are fairly close together, is:

> **the HCF will also be a factor of the sum and of the difference of the numbers.**

This comes under the Sutra *By Addition and By Subtraction*.

EXAMPLE 10

Find the HCF of 411 and 417.

This means the HCF will divide into not just 411 and 417
but also 411 + 417 = **828** and
417 − 411 = **6**.

This 6 is particularly useful. It tells us that the HCF is either 6 or a factor of 6 (i.e. 6 or 3 or 2 or 1).

[You can see that no number higher than 6 will divide into 411 and 417, because if, say, 7 were to divide that would mean that they were both multiples of 7 and so they would have to differ by at least 7. Similarly for any other number higher than 6]

It is easy to see that 2 is not a common factor (the last figures are not both even) and that 3 is a common factor. So 3 is the highest common factor.

 EXAMPLE 11

Find the HCF of 90 and 102.

The difference here is 12, so possible common factors are 12, 6, 4, 3, 2, 1, that is, 12 and all its factors.
3 is seen to be a common factor of 90 and 102,
and 2 is also, but not 4.

Therefore 2×3 = 6 is the HCF.

This method also indicates that if the numbers are consecutive, like say, 64, 65, then only 1 will divide into both and so they must always be relatively prime.

 EXERCISE 5

Find the HCF:

a 56, 58	**b** 51, 54	**c** 52, 55	**d** 77, 79
e 72, 78	**f** 85, 100	**g** 57, 69	**h** 91, 98
i 55, 75	**j** 56, 66	**k** 38, 49	**l** 27, 28
m 207, 213	**n** 530, 534	**o** 322, 343	**p** 513, 531

"When he was twelve George Parker Bidder (1806-1878) of Devonshire, England was asked 'If a pendulum clock vibrates the distance of 9.75 inches in a second of time, how many inches will it vibrate in 7 years, 14 days, 2 hours, 1 minute, 56 seconds, each year being 365 days, 5 hours, 48 minutes, 55 seconds?' In less than one minute he gave the answer, 2,165,625,744.75 inches.

22

PROPORTIONATELY

If we want to bake a cake or mix mortar for laying bricks we must mix the ingredients in the right **proportions**, or we might say, in the right **ratio**.

To make mortar we can mix sand and cement in the proportions 3 parts of sand to 1 part of cement.
We can write this proportion, or ratio, as **3 : 1**.

So you can mix 3 shovelfulls of sand with 1 shovelfull of cement and (when you add the water) you will get good mortar.

EQUAL RATIOS

But if you want a lot of mortar you may mix 6 shovels of sand with 2 shovels of cement.
The ratio is still the same, you have still got 3 times as much sand as cement.

So we can write   3 : 1 = 6 : 2   these are equal ratios.

Similarly   3 : 1 = 9 : 3   the first number is still 3 times the second.

1 Which of the ratios below is the same as 3 : 1?

$$12:4 \quad 30:10 \quad 6:3 \quad 24:8 \quad 18:12 \quad 5:15$$

6:3 and 18:12 are not the same, as the first is not 3 times the second.
5:15 is the not same because it is the second which is 3 times the first.
12:4, 30:10 and 24:8 are equal to 3:1.

 EXAMPLE 1

Other examples of equal ratios are:

$8:4 = 2:1,$ $2:3 = 4:6,$ $3:3 = 2:2,$ $0.5:1 = 1:2 = 5:10,$ $17:8 = 34:16.$

---

2 You will need Worksheet 2 and some coloured pencils.
In each of the 8 questions circle all the ratios which are equal with the same colour.
So in question 1 if you circle 1:2 in red, for example, you should circle all ratios which are equal to 1:2 in red also.
You will need 4 colours for the first question because there are 4 different ratios in it.

SIMPLIFYING RATIOS

The ratios 4:1 and 24:6 are equal but 4:1 is simpler because the numbers are small and whole.
Of all the ratios equal to 4:1 the ratio 4:1 is the simplest.

Every ratio which is not a simplest ratio can be simplified to a simplest ratio.

 EXAMPLE 2

24:4 is equal to many other ratios, but the simplest is 6:1.
We divide the ratio by the biggest number we can.

> That is to say, we divide the numbers in the ratio by their Highest Common Factor.

In this case the HCF of 24 and 4 is 4.
So we divide 24 and 4 by 4 to get 6:1.

 EXAMPLE 3

Similarly we can simplify the following ratios:

$18:9 = \mathbf{2:1}$ we divide by 9 here,

$20:15 = \mathbf{4:3}$ we divide by 5,

$36:12 = \mathbf{3:1}$ the HCF is 12,

$100:30 = \mathbf{10:3},$

$1\frac{1}{2} : 2\frac{1}{2} = \mathbf{3:5}$ here we multiply $1\frac{1}{2}$ and $2\frac{1}{2}$ by 2 to get whole numbers.

EXAMPLE 4

If units are involved we have to be careful.
A ratio of 1 minute to 1 hour is not a ratio of 1:1 but of 1:60.
And a ratio of 3 days to 2 weeks is a ratio of 3:14.

EXERCISE 1

Write the following ratios in their simplest form:

a 9:3	**b** 5:15	**c** 21:14	**d** 14:10	**e** 8:12	**f** 8:16
g 12:2	**h** 12:3	**i** 12:4	**j** 12:5	**k** 12:6	**l** 12:12
m 33:77	**n** 60:20	**o** 80:2	**p** 28:16	**q** 54:4	**r** 200:700
s 15:27	**t** 20:36	**u** $2\frac{1}{2}$:3	**v** $4\frac{1}{2}$:$2\frac{1}{2}$	**w** 1.5:4.5	**x** 2:$\frac{1}{2}$

y 3 weeks : 5 days

z 2 hours : 1 hour 45 minutes

FINDING EQUAL RATIOS

Instead of simplifying ratios we may, for some purposes, need to do the opposite.

EXAMPLE 5

Give three ratios the same as 5:2.

We can multiply, or divide these numbers by any number we like.
So there are many answers to this question.
If we choose to multiply by 3, 4 and 5 we get <u>15:6, 20:8, 25:10</u>.

3 Give three ratios the same as 2:3.

4 Give 5 ratios the same as 12:4.

EXAMPLE 6

15:5 = x:6   what is the value of x?

Here we have two equal ratios, but one of the numbers is unknown.
We notice that 15 is three times bigger than 5,
so x must be three times bigger than 6.
So <u>x = 18</u>.

EXAMPLE 7

7:3 = x:6   what is the value of x?

Here the first ratio, 7:3, is not such an easy one.
But looking at the last number in each ratio (3 and 6) we see that 6 is double 3.
So x must be double 7, which is 14, x = 14.

 EXAMPLE 8

x:5 = 4:10   5 is half of 10, so x = 2,

6:x = 2:7   6 is three times 2, so x = 21,

8:1 = 24:x   x = 3.

 EXERCISE 2

Find x in each of the following:

a 1:3 = x:6   **b** 2:10 = x:30   **c** 2:3 = x:9   **d** 3:7 = x:21

e x:12 = 5:4   **f** x:20 = 7:5   **g** 1:x = 4:12   **h** 2:x = 4:6

i 3:2 = 15:x   **j** 7:6 = 21:x   **k** 4:x = 2:1   **l** 50:2 = x:1

m 10:3 = 5:x   **n** 12:6 = 3:x

RATIO PROBLEMS

Many common problems lead to ratio equations like the ones above.

 EXAMPLE 9

If 12 identical pencils cost 60p, what will be the cost of 3 pencils?

This is the same as solving 12:60 = 3:x.
The solution is x = 15 so 3 pencils will cost 15p.

 EXERCISE 3

Solve the following problems:

a If 6 pens cost £12, what will 4 such pens cost?

b If 5 apples cost 120p, what will be the cost of 15 apples?

c If 20 pears cost £2.40, what will 5 cost?

d If 4 identical bicycles cost £300, what will 20 of these cost?

SPLITTING IN A RATIO

There is another kind of problem which is solved using ratios.

 EXAMPLE 10

Jane and John are given £20 to share between them, but Jane must get 3 times as much a John. What do they each get?

You have had problems like this before.
The problem could be restated as: **split 20 in the ratio 3:1**.
So two numbers add up to 20 and one is 3 times the other.
The numbers must be 15 and 5: Jane gets £15 and John gets £5.

 EXAMPLE 11

Jack and Jill share £20 in the ratio 2:3. What do they each get?

You may see the answer straight away.
If not, add up the numbers in the ratio: 2 + 3 = 5.
This tells you that for every £5 shared, £2 goes to Jack and £3 goes to Jill.
So how many £5 are there in £20? There are four.
So Jack gets £2 four times, and Jill gets £3 four times.
Then Jack will get £8, and Jill will get £12.

Notice that £8 and £12 are in the ratio 2:3, and that £8 + £12 = £20.

 EXERCISE 4

Split the following numbers in the given ratio. If the answer is obvious just write it down (and check it works), if not use the method shown in Example 11 above.

a 15 in 2:1	**b** 10 in 4:1	**c** 12 in 1:3	**d** 8 in 3:1
e 18 in 5:1	**f** 22 in 10:1	**g** 21 in 1:2	**h** 25 in 1:4
i 15 in 2:3	**j** 25 in 3:2	**k** 14 in 5:2	**l** 14 in 3:4
m 60 in 1:5	**n** 60 in 1:4	**o** 60 in 1:3	**p** 60 in 1:2

EXTENDED RATIOS

A ratio may consist of more than two numbers.
You may, for example, see 1:2:8 or 7:3:2:5.

1:2:8 means there are three things in which the second is double the first and the last is 8 times the first.

It also shows that the last is 4 times the second (because of the ratio 2:8).

A recipe for pastry is   100g of flour,
                            50g of oil,
                            25g of water.

Of course we may want to double these amounts, or halve them etc.
But the ratio must always be   100:50:25.
Dividing this through by the HCF, that is 25, we can write 100:50:25 = 4:2:1.

In other words there must be twice as much oil as water and twice as much flour as oil.
Or if you like, 4 times as much flour as water.

 EXAMPLE 12

A ratio could have mixed units.
Here is a recipe for chocolate mousse: 2 cups of heavy cream,
                                            4 ounces of baking chocolate,
                                            1 teaspoon of vanilla.

You could still double everything if you wanted to provided you use the right units (cups ounces, teaspoons).

 EXERCISE 5

Simplify the following ratios:

a 2:4:6   **b** 10:2:8   **c** 9:6:3   **d** 8:6:4:2   **e** 15:10:15   **f** 12:18:24

g 5:5:15:20   **h** 4:8:8:6:2   **i** 60:20:40   **j** $4:3:\frac{1}{2}$ **k** $1\frac{1}{2}:7:2\frac{1}{2}$ **l** $1:3:\frac{1}{4}$

m 1kg:500g:250g   **n** 6 cups : 8 ounces : 2 teaspoons

 EXAMPLE 13

Split £60 in the ratio 5:3:4.

Since 5+3+4 = 12, and 60 ÷ 12 = 5, we multiply the ratio by 5.

5×5=25, 3×5=15, 4×5=20.
So the answer is £25, £15, £20 (these can be done mentally after a little practice).

 EXERCISE 6

Split the following numbers in the given ratio:

a 12 in 3:2:1   **b** 30 in 2:7:1   **c** 24 in 3:4:5   **d** 20 in 1:1:2
e 40 in 4:1:2:3   **f** 100 in 5:10:5   **g** £1.20 in 1:2:17   **h** £1 in 2:3:5
i 72 in 3:1:3:1

23

BY ONE MORE THAN THE ONE BEFORE

The Vedic Sutras give us some very fast methods of calculation.
In particular the Sutra Ekadhikena Purvena, *By One More than the One Before* gives us some special multiplication devices which are extremely efficient.

SQUARING NUMBERS THAT END IN 5

Squaring is multiplication in which a number is multiplied by itself:
so 75×75 is called "75 squared" and is written 75^2.
The formula *By One More Than the One Before* provides a beautifully simple way of squaring numbers that end in 5.

 EXAMPLE 1

In the case of 75^2, we simply multiply the 7 (the number before the 5) by the next number up, 8. This gives us 56 as the first part of the answer, and the last part is simply 25 (5^2).

So $75^2 = \underline{56/25}$ where $56 = 7 \times 8$, $25 = 5^2$.

 EXAMPLE 2

Similarly $65^2 = \underline{4225}$ $42 = 6 \times 7$, $25 = 5^2$.

 EXAMPLE 3

And $25^2 = \underline{625}$ where $6 = 2 \times 3$.

EXAMPLE 4

Also since $4\frac{1}{2} = 4.5$, the same method applies to squaring numbers ending in $\frac{1}{2}$.
So $4\frac{1}{2}^2 = \underline{20\frac{1}{4}}$ where $20 = 4\times5$ and $\frac{1}{4} = \frac{1}{2}^2$.

The method can be applied to numbers of any size:

 EXAMPLE 5

$305^2 = \underline{93025}$ where $930 = 30\times31$

Even for large numbers like, say, 635, it is still easier to multiply 63 by 64 and put 25 on the end than to multiply 635 by 635.

 EXERCISE 1

Square the following numbers:

a	55	b	15	c	$8\frac{1}{2}$	d	95	e	105

f	195	g	155	h	245	i	35	j	$20\frac{1}{2}$

k 8005   **l** 350   **m** What number, when squared, gives 2025?

MULTIPLYING NUMBERS WHOSE FIRST FIGURES ARE THE SAME AND WHOSE LAST FIGURES ADD UP TO 10, 100 ETC.

 EXAMPLE 6

Suppose we want to find 43×47 in which both numbers begin with 4 and the last figures (3 and 7) add up to 10.

The method is just the same as in the previous section.
Multiply 4 by the number *One More*: $4 \times 5 = 20$.

Then simply multiply the last figures together: $3 \times 7 = 21$.

So $\underline{43 \times 47 = 2021}$ where $20 = 4\times5$, $21 = 3\times7$.

 EXAMPLE 7

Similarly $\underline{62 \times 68 = 4216}$ where $42 = 6\times7$, $16 = 2\times8$.

EXAMPLE 8

Find 204 × 206

Here both numbers start with 20, and 4 + 6 = 10, so the method applies.

204 × 206 = 42024   (420 = 20×21, 24 = 4×6)

 EXAMPLE 9

93 × 39   may not look like it comes under this particular type of sum,
but remembering the *Proportionately* formula we notice that 93 = 3×31,
and 31 × 39 does come under this type:

31 × 39 = 1209   (we put 09 as we need double figures here)

so   93 × 39 = 3627   (multiply 1209 by 3)

The thing to notice is that the 39 needs a 31 for the method to work here: and then we spot
that 93 is 3×31.

 EXAMPLE 10

Finally, consider 397 × 303.

Only the 3 at the beginning of each number is the same, but the rest of the numbers (97 and
03) add up to 100.

So again the method applies, but this time we must expect to have four figures on the right-
hand side:

397 × 303 = 120291   where   12 = 3×4, 0291 = 97×3

? **EXERCISE 2**

Multiply the following:

a 73 × 77	**b** 58 × 52	**c** 81 × 89	**d** 104 × 106
e 42 × 48	**f** 34 × 36	**g** 93 × 97	**h** 27 × 23
i 297 × 293	**j** 303 × 307	**k** 64 × 38	**l** 88 × 46
m 33 × 74	**n** 66 × 28	**o** 36 × 78	**p** 46 × 54
q 298 × 202	**r** 391 × 309	**s** 795 × 705	**t** 401 × 499
u 802 × 499			

ROUNDING

There are three types of rounding:

> **A:** rounding to the nearest 10, 100, 1000 etc.
> **B:** significant figures.
> **C:** decimal places.

A: Sometimes we need to round a number to the nearest 10, 100, 1000 etc.

For example, a broadcaster may not want to report the exact number of spectators at a football match, even if the exact number is available. It may be given to the nearest thousand.

 EXAMPLE 11

Round to the nearest 10: **a** 4,568   **b** 634   **c** 625   **d** 3,399

a there is a 6 in the tens position of 4568 and since the next figure is 8 the 6 is rounded up to 7 and we write: 4,570.

> To round to the nearest 10. Look at the figure in the tens position:
>   if the figure after it is **5 or more**, **increase** the tens figure **by 1** and put a zero after it,
>   if the figure after it is **less than 5, leave it as it is** and put a zero after it.

In this example we could just say 68 is 70 to the nearest 10, so we replace 68 by 70.

b For 634 we see a 3 in the tens position so we leave it but replace the 4 with a 0.
So 634 rounds to 630.

c 625 has a 2 in the tens position but since it is followed by 5 the 2 is increased by 1 to 3: 630.

d For 3,399 we must increase the 9 in the tens place by 1 as it is followed by 9.
So we get 3,400. (since 339+1 = 340)

 EXAMPLE 12

Round **a** 2,468 to the nearest 100   **b** 54,321 to the nearest 1000.

a Here we apply the same method but look at the 100's figure and the figure that follows it.
The hundreds figure of 2,468 is 4 and is followed by a 6 so the 4 is increased by 1 and we put 2 zeros at the end giving 2,500.

b the 4 here in the thousands position in 54,321 is followed by 3 so the 4 remains as it is with 3 zeros after it: 54,000.

 EXERCISE 3

Complete TABLE A on Worksheet 3 (some have been done for you).

Note: You only need to look at the appropriate figure and the next one. If you want a number to the nearest 1000, for example, look at the 1000's and the 100's, not the 10's or units.

B: Significant Figures (S.F.)

A scientist needs to always state how accurate the data or results of an experiment are and will quote the number of significant figures for those numbers.

The most significant figures of a number are the figures on the left.
So if we want a number to two significant figures (often written 2 S.F.) we are interested only in the first two figures of the number, and the rest are replaced with zeros.

> We use the same rule as used in the previous section: that if the last significant figure is followed by a figure of 5 or more the last significant figure is increased by 1.

 EXAMPLE 13

Find   13 579   to **a** 1 significant figure   **b** 2 significant figures
                         c 3 significant figures   **d** 4 significant figures.

a   1 S. F. means the first (on the left). The 1 is followed by 3 so 1 is unchanged and the other figures are replaced with noughts: <u>10 000</u>.

b   2 S.F. means the first two figures: 13. This is followed by a 5 so this 13 becomes 14, and is followed by 3 noughts: <u>14 000</u>.

c   For 3 S.F, we have 135 followed by 7, so 135 becomes 136, and 2 noughts are added: <u>13 600</u>.

d   For 4 S.F. we get <u>13 580</u>.

 EXAMPLE 14

Find 0.05037   **a** to 3 S.F.   **b** to 2 S.F.

a   0.05037 becomes <u>0.0504 to 3 S.F.</u> because 5 is the first significant figure here- **the 0's before the 5 are not counted, but the 0 within the number is counted**.

b   0.05037 becomes <u>0.050 to 2 S.F.</u> You may be tempted to leave the last 0 off and write 0.05, but you should not do this because 0.050 indicates that the number is correct to 2 S.F. whereas 0.05 looks as though it has only 1 S.F.

 EXERCISE 4

Complete TABLE B on your Worksheet.

C: Decimal Places (D.P.)

This is similar to the above and is used where there are decimal places in a number.
We say the number is correct to so many decimal places.

> So instead of counting the figures in a number from the left-hand end
> as in the previous section, we count from the decimal point.

 EXAMPLE 15

Write the number 54.1843 correct to   **a** 1 decimal place   **b** 2 decimal places.

a 54.1843 to 1 decimal place is 54.2. Up to one decimal place we see 54.1, but as this 1 is
followed by a large figure, 8, the 1 is increased to 2.

b 54.1843 to 2 D.P. is 54.18. The 8 in the second decimal place is followed by a small
figure, 4, so we just leave it off.

 EXERCISE 5

Complete TABLE C on your Worksheet.

 ACCUMULATOR GAME   for 2, 3 or 4 players.

A game for 2 to 5 players. You need only a pair of dice. The only writing that is allowed is for
keeping scores.

1) One player throws the dice. The two numbers obtained are then multiplied together and the
person who threw the dice gives their product (so if 4 and 5 are thrown the player who threw
the dice should say '20'.

2) If the players agree that the answer is correct the player who threw the dice is given a point.
If a player thinks the answer given is not correct he/she should say so and the first player to say
an incorrect answer is incorrect gets a point, provided they can give the correct answer. If
however the player who says the answer is incorrect is wrong everyone else gets a point. If two
people say an answer is incorrect at the same time they can both have a point if they are right.

3) The dice are now thrown by the person to the left of the previous thrower but this time the
product is added to the last answer. So if 4 and 5 are thrown the first time and then 3 and 3 the
answer should be 29 (20 + 9). The thrower gives their total.

Steps 2 and 3 above are then repeated continually until the accumulated total reaches or
exceeds 200 in which case the players finish the last part of the game and the player with the
most points is the winner.

*"Algebra is generous;
she often gives more than is asked of her."*

D'Alembert (1717-83)

FRENCH MATHEMATICIAN

24

ALGEBRA

Algebra is one of the main branches of the mathematics and deals with using symbols to represent things. For example your name is a symbol which represents yourself.

In algebra we often use the letters of the alphabet as symbols to represent whatever we like. So you could say "x represents the number of children in a class" even though you don't say which class you are talking about. You, or someone else gives the symbol its meaning.

Sometimes we have to work with these symbols without knowing what they stand for.

USING LETTERS

Suppose we see, for example, $2a + 3a = 5a$.
This tells us that, whatever "a" is, if 2 of them are added to 3 of them, there will be 5 of them altogether.

> So $5a$ means $5 \times a$ which means $a + a + a + a + a$.

Similarly $7b - 3b = 4b$.
This is like saying if there are 7 bananas and we take 3 of them away, there will be 4 bananas left.

We might have a mixture of letters in the same sum.

EXAMPLE 1

$3a + 4b + 2a + 6b = \underline{5a + 10b}$

we can think that 3 apples and 4 bananas are put with 2 apples and 6 bananas, giving apples and 10 bananas altogether.

When we see $3a + 4b + 2a + 6b$ we see four things added up.
We call something like $3a + 4b + 2a + 6b$ an **expression**, and this particular expression has four **terms**, since four things are added up.

Notice that $5a + 10b$ is the answer, we cannot add the 5a and the 10b together.

Another useful word is **coefficient** which is the number in front of a letter: so the coefficient of 5a is 5, and the coefficient of 7abc is 7.

 EXAMPLE 2

$8a + 4a + 5b + a = \underline{13a + 5b}$

Note here that "a" means "1a", the coefficient is 1.
So we add $8+4+1$ to get the 13a.

 EXAMPLE 3

$12a + 5 + 50a - 2 = \underline{62a + 3}$

here we have a mixture of "a" terms and "unit" terms:
for the "a" terms we find 12+50,
and for the units we find 5–2.

 EXERCISE 1

Simplify the following expressions:

a $4a + 5a$ **b** $7b - 5b$ **c** $7c + c$ **d** $2a + 11a - 4a - a$

e $3a + 2b + 4a + 6b$ **f** $9a + 21b + a - 7a$ **g** $7m + 3n + 9m - 2n + 60m$

h $8a + 5 + 4a + 9$ **i** $4x + 7 - 2x - 4$ **j** $a + 5b + 7 + 8b + 2a + 7$

k $7p + 3q + p - 3q$ **l** $x + y + x - y$ **m** $11a + 20y - 3a - 2y + a - 8y$

We have seen above that we can only combine terms of the same kind: we can add or subtract "a" or "b" terms for example, but we cannot combine the "a" and "b" terms together.
So if the answer is $2a + 3b$ we cannot add 2a and 3b because they are different types of term.

EXAMPLE 4

$3x^2 + 5x^2$ means that we have x^2 three times, and also x^2 five times:

so altogether we have $\underline{8x^2}$, we just add the coefficients.

 EXAMPLE 5

Similarly in $3ab + 4bc + 5ab - bc$ we can combine the "ab" terms and the "bc" terms, because they are the same type, and the answer is $\underline{8ab + 3bc}$.

 EXAMPLE 6

Simplify $4x^2 + 6x + 9 + 8x^2 + 5x + 1$

Here there are three different types of term: x^2, x and units.
So the answer is $\underline{12x^2 + 11x + 10}$.

 EXERCISE 2

Simplify:

a $7x^2 + 11x^2$

b $20y^2 - 9y^2$

c $5ab + 6ab$

d $5ab + 8cd + 6ab - 3cd$

e $4pq + 3qr + 6qr + 5pq$

f $3ab + 4bc + 5cd - 2ab + bc + 7cd$

g $xy + 3 + 8xy + 30$

h $3x^2 + 5 + 4x^2 - 2$

i $4a^2 + 7b^2 + 5a + 2b^2 - 4a$

j $ax + by + ax + 2by$

k $8x^2 + 9x + 10 + x^2 - 4x$

l $ab + b + a + ab + 3a - b$

BRACKETS

 EXAMPLE 7

Expand $3(4a + 5b)$.

Since we know that $3x$ means $x + x + x$ it follows that
$3(4a + 5b)$ means $4a+5b + 4a+5b + 4a+5b = 12a + 15b$.
So $3(4a + 5b) = \underline{12a + 15b}$.

The easiest way to get the answer is to multiply 4a by 3 and also multiply 5b by 3.

EXAMPLE 8

Expand and simplify $4(2x + 5) + 5(3x + 2)$.

Multiplying the contents of the first bracket by 4 and the second bracket by 5 we get
$8x+20 + 15x+10$,
and this simplifies to $\underline{23x + 30}$, which is the answer.

EXAMPLE 9

Find ½(6A + 8B – 10) – 2B.

Here we halve the terms in the brackets,
so we get 3A + 4B – 5 – 2B which is <u>3A + 2B – 5</u>.

 EXERCISE 3

Simplify:

a 5(4a + 5b)

b 2(7x + 3y) + 4x

c 8(5a + 5b) +3(3a + 4b)

d 20(7x – 9)

e 4(17a + 23b – 70c)

f 3(15p + 8q) + 5(p – 3q)

g 9(a + b) + 3(a – b)

h 20(30a + 40b + 50) + 6a

i ½(4x + 6y)

j Find 3(7 + 4) by   **i** adding up the bracket first and then multiplying by 3,
                            ii multiplying the bracket out first and then adding up.

You should get the same answer both ways.

k Find 4(18 + 60)

l Find 6(18 – 16) + 3(7 + 4)

FACTORISING

 EXAMPLE 10

Fill in the brackets: 4m+6n+20 = 2(   ).

This is the reverse of multiplying brackets out.
The bracket, when multiplied by 2 gives 4m+6n+20,
so the bracket must contain <u>2m+3n+10</u>

 EXAMPLE 11

Factorise 15a+18b.

Here we first have to find the largest number that divides into both 15 and 18.
This is 3, the **highest common factor** of 15 and 18.

So we can say that 15a+18b = 3(   ) and then fill in the bracket.
15a+18b = <u>3(5a+6b)</u> .

This process is called **factorisation**.

EXAMPLE 12

Factorise 24x+12y–42.

The highest common factor of 24 , 12 and 42 is 6, so we write 24x+12y–42 = 6(   ).

And then we can see that 24x+12y–42 = 6(4x+2y–7)

❊ Check this answer is correct by multiplying the bracket out.

 EXERCISE 4

Copy out and fill in the brackets:

a 8p+12q = 2(   )   **b** 8p+12q = 4(   )   **c** 9x–15y = 3(   )

d 21a+14b+7 = 7(   )   **e** 54a+36b–18c = 6(   )   **f** 52x+39y = 13(   )

Factorise the following (take out the largest possible factor in each case):

g 8a+14b   **h** 9a+12b   **i** 9a–3   **j** 20x+15y   **k** 8x+64

l 4a+6b+8c   **m** 50a–20b   **n** 18p+12q   **o** 24a+16b–40c   **p** 18–9x

q 7+7y

 EXAMPLE 13

Simplify and factorise 2(3a+6b) + 3(4a–b)

Here we multiply out the brackets, then simplify and then factorise.
2(3a+6b) + 3(4a–b) = 6a+12b + 12a–3b = 18a+9b = 9(2a+b)

 EXERCISE 5

Simplify and factorise:

a 3(3a+5b) + 5(a–b)   **b** 2(12x+2y) + 3(3x+6y)   **c** 6(2x+y) +2(x+3y)

d 6(2x+3y) + 3(x–2y)   **e** 5(2x+4y) + 2(2x+3y)   **f** 2(2x+y) + 3(3x+y) + 2x

SUBSTITUTION

Although it is not possible to add up something like 5a + 10b, if the values of a and b are known, the value of 5a + 10b can be found.

EXAMPLE 14

If $a = 3$ and $b = 4$, find the value of $5a + 10b$

If $a = 3$ then $5a = 15$, and if $b = 4$ then $10b = 40$.
So $5a + 10b = 15 + 40 = \underline{55}$.

The process of substitution comes under the Vedic formula *Specific and General* because a letter, which has no particular value, is being replaced by a particular value.

Where several calculations have to be made, as in the example below, it is important to follow the standard conventions on the order in which the calculations are done.

 EXAMPLE 15

If $x = 20$ and $y = 3$ find   **a** $x - y + xy$ **b** $y(2x+1)$ **c** $2x^2$ **d** $(2x)^2$

> The standard convention is that **brackets** are worked out first;
> then **multiplication and division**
> and finally **addition and subtraction.**

Replacing x with 20, and y with 3 (this is called substituting):

a $x - y + xy = 20 - 3 + 20 \times 3 = 20 - 3 + 60 = \underline{77}$ note we first multiply 20 by 3 **before** doing the addition and subtraction.

b $y(2x + 1) = 3(40 + 1)$ since 2x is 40
         $= 120 + 3 = \underline{123}$ note that we work out the bracket first.

c $2x^2 = 2 \times 20^2 = 2 \times 400 = \underline{800}$.

d $(2x)^2 = 40^2 = \underline{1600}$.

> Note carefully the difference between $2x^2$ and $(2x)^2$.

 EXERCISE 6

Given that $p = 30$ $q = 5$ and $r = 7$ find:

a $p + q + r$ **b** $p - q + r$ **c** $2p + 3q - 4r$ **d** $3(7q - 1)$ **e** $p^2 + q^2$

f $100 - 14q$ **g** $pr + qr$ **h** $3pq$ **i** $pqr - 1$ **j** $2(p + 3q - 2r + 1)$

k $p^2 - q^2 - r^2$ **l** $2p^2$ **m** $(2p)^2$ **n** $(13 + r)^2$ **o** $(3q^2 + 5)^2$

p $9q^2 + 8q - 7$ **q** $(p + q)(r - q)$ **r** $(4q - 16)(3p + 1)$ **s** $(2p + 3(q - 1) + 4)$

MULTIPLE SUBSTITUTIONS

Sometimes it is useful to substitute a series of values into an algebraic expression.

 EXAMPLE 16

Write out the values of $2n+1$ as n takes the values $n=1,2,3,4\ldots$ etc.

When $n=1$ $2n+1 = 3$,
when $n=2$ $2n+1 = 5$,
when $n=3$ $2n+1 = 7$, and so on.

We see that the odd numbers are generated, starting at 3.
So the answer is $3, 5, 7, 9 \ldots$

EXERCISE 7

Put $n=1,2,3,4,5$ in the following:

a $2n$ **b** $n + 5$ **c** $2n - 1$ **d** $3n - 2$ **e** $20 - 2n$ **f** $100 - 5n$

g n^2 **h** $(n + 1)^2$ **i** $3n^2$ **j** $(3n)^2$ **k** n^3 **l** $n^2 + 2n + 1$

m $\frac{1}{2}n(n + 1)$

Note that in question **a** above **2n** gives the **even numbers**,

and that in question **m** $\frac{1}{2}\mathbf{n(n+1)}$ gives the **triangle numbers**.

1 What kind of numbers are generated in **c** above?

In **a** to **f** above **the coefficient of n gives the increase from one term to the next**.
In **d** for example where the coefficient of n is 3 the numbers are increasing by 3.

2 What do you think the sequence generated by **7n+3** would increase by from one term to the next?

"Vedic Mathematics is the science and technology of precision and order, which maintains the integrity of Unity and at the same time spreads diversity within it."

Maharishi Mahesh Yogi

FOUNDER OF TRANSCENDENTAL MEDITATION

25

AREA

Area is the amount of space inside a shape. It is measured in square units: square kilometres, square metres, square inches, square centimetres etc.

RECTANGLES AND SQUARES

For a symmetrical shape like a rectangle it is easy to find the area.

➡ EXAMPLE 1

What is the area of the rectangle below?

3cm

6cm

If the rectangle was filled with square centimetres (sq.cm.) there would be 6 of them in the bottom row, and since there would be 3 rows of 6 there would be 18 altogether.
So the area is exactly <u>18 sq.cm.</u>

We clearly find the area of such a shape by multiplying the base of the rectangle by its height: **Area of a rectangle = base × height**.

And this formula is true for all rectangles, even if the sizes are not whole numbers

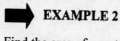 **EXAMPLE 2**

Find the area of a rectangle 7mm by 4mm.

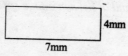

Base × height = 7 × 4 = 28mm² (28 square millimetres).

 EXERCISE 1

Find the areas of the rectangles with the given sizes:

a 7cm by 30cm   **b** 2½cm by 8cm   **c** 2.3m by 30m

d 48in by 42in   **e** 300m by 7m   **f** 3.5km by 8km

A square is even easier since the height is equal to the base:

Area of a square = base squared.

 EXAMPLE 3

Find the area of a square with a base length of 40m.
Area = 40^2 = 1600m².

EXAMPLE 4

Find the area of a rectangle 30cm by 2m.

Here we must be careful as we must give the answer in square centimetres or square metres. Since 2m = 200cm we can give the area as 30 × 200 = 6000 sq.cm.
This could also be written as 6000 cm².

 EXERCISE 2

Find the areas of the rectangles with the given sizes (where different units are involved it is usually best to use the smaller unit):

a 7cm by 2m   **b** 2½m by 2cm   **c** 2m by 30cm   **d** 300m by 7km

e 3.5m by 8cm   **f** 3 feet by 3 inches

Find the area of a square of side **g** 5km   **h** 65cm

IRREGULAR SHAPES

Finding the area of an irregular shape like a leaf is not so straightforward: we can only expect to get an approximate value for the number of square centimetres in it.

1 Place your hand on a sheet of centimetre squared paper and draw round it.
   You are going to estimate the area of your hand by counting the square centimetres inside it.

   Put a dot in each complete square to ensure you do not count it twice.
   But what do we do with the squares which are not complete?
   We count each one as a whole square if it is greater than half a square, otherwise we do not count it at all.

   Find the area of your hand.

✳ Choose some other irregular shapes such as leaves and find their area in this way.

Though it is not possible to find an exact value for the area of an irregular shape like a island, it is possible to get a good estimate.

  EXAMPLE 5

Find the area of the garden shown

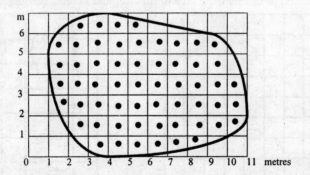

The method we use is to count up the whole squares inside the shape, and the parts which are not whole squares are counted as a whole squares if they are more than half covered by the garden.

We find 57 such squares and so the area of the garden is estimated as 57m².

A scientist is investigating the rainfall on some remote islands and needs to know the size or area of the islands. The scientist has some maps of the region and can estimate the areas from them.

EXAMPLE 6

The diagram below shows a map of an island on a grid.
Estimate the area of the island.

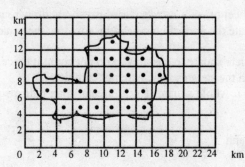

Using the same method as before we count the total number of dots to be 23.

But the area is not 23 square kilometres because the grid is drawn every 2 kilometres.
This means the area of each square on the grid is 4 square kilometres, and so the area
of the island is approximately 4 × 23 = <u>92 square kilometres</u>.

EXERCISE 3

Use the above method to estimate the area of the irregular objects below:

a
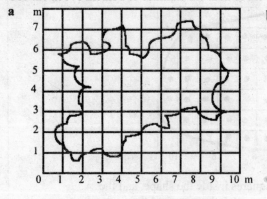

b
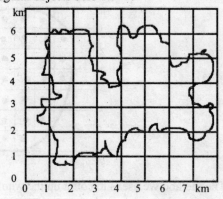

c
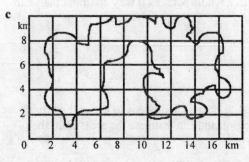

d
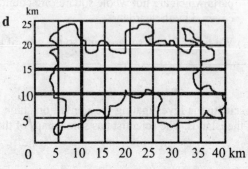

COMPOSITE SHAPES

Shapes which can be broken down into rectangles can also be measured.

 EXAMPLE 7

Find the area of the shape below:

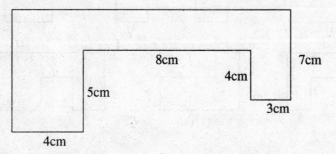

We can extend the vertical 4cm and 5cm lines to form three rectangles A, B, C:

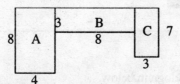

✳ Check that you agree with the figures on this diagram.

Applying the Sutra Gunitasamuccaya Samuccayagunita, *The area of the whole is the sum of the areas* we get Area = 4 × 8 + 8 × 3 + 3 × 7 = 77 cm².

The above shape could also be split up by extending the horizontal 8cm line.

2 Show that the total area obtained in this way is also 77 cm².

It is sometimes useful to use the Sutra *By the Completion or Non-Completion* in these problems.

In the above example we could enclose the figure in a rectangle, find its area, and subtract the unwanted part.

3 Find the area by this method.

? **EXERCISE 4**

Find the areas (assume all lengths are cm):

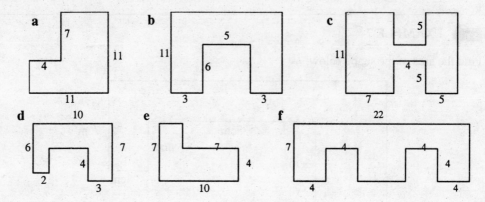

PARALLELOGRAMS

➡ EXAMPLE 8

Find the area of the parallelogram below.

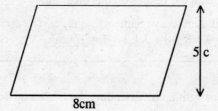

We can apply the formula *By Addition and By Subtraction* here and move the triangle on the left to the right-hand side of the figure:

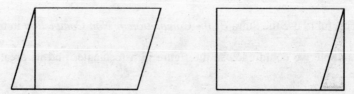

Since the area of the rectangle on the right is the same as the parallelogram on the left, and since the base and height of the rectangle are 8cm and 5cm, the area of both figures is 8×5 = 40cm².

Since this can be done for any parallelogram we can use the formula:

> **Area of a parallelogram = base × perpendicular height.**

It is important to emphasise that the height is perpendicular because if we had a parallelogram like this:

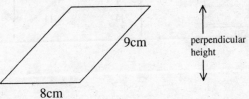

the area would not be 8×9. We need to know the perpendicular height, which is not given here.

Sometimes it is useful to think that the base of a parallelogram is not at the bottom of the figure.

EXAMPLE 9

Find the area of the parallelogram:

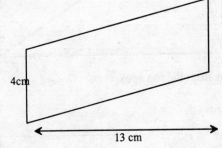

Here it is best to think of the 4cm line as the base of the parallelogram.
Then 13cm will be the perpendicular height.
So the area is 4×13 = <u>52 cm²</u>

 EXERCISE 5

Find the areas (all units are m):

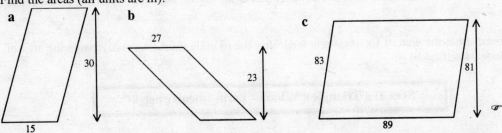

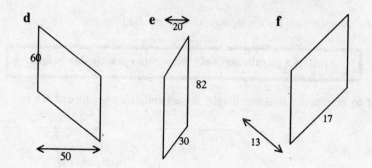

TRIANGLES

Suppose that we have a triangle and we complete the rectangle which encloses the triangle:

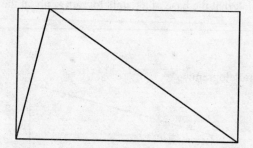

Next we drop a perpendicular from the top apex:

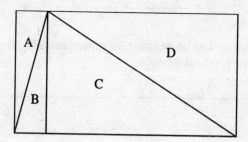

We can see from the diagram that the areas A and B are equal, and that the areas C and D are equal.

This means that the area of the rectangle enclosing the triangle must be exactly twice the area of the triangle itself. And so:

> **Area of a Triangle = ½ base × perpendicular height.**

EXERCISE 6

Find the area of the triangles below:

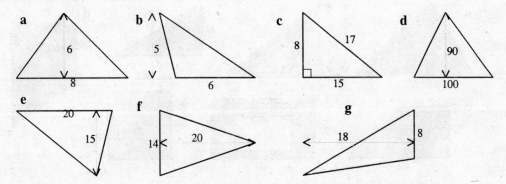

This means we can now find areas of triangles and also of shapes which can be split up into triangles, rectangles or parallelograms.

➡ **EXAMPLE 10**

Find the area of the figure below:

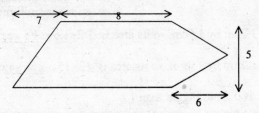

We can split this into two triangles and a rectangle:

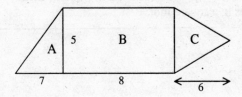

The area of A is ½ × 7 × 5 = 17½,
the area of B is 8 × 5 = 40,
the area of C is ½ × 5 × 6 = 15.   (think of the side of length 5 as the base, so that 6 is the height)
And so the total area is 17½ + 40 + 15 = 72½ square units.

? **EXERCISE 7**

Find the shaded areas below (all units are m):

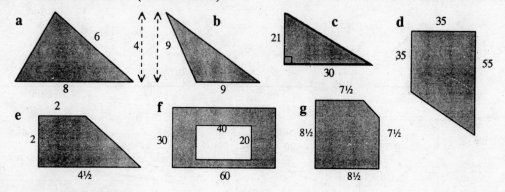

UNITS OF AREA

Care needs to be taken when two or more units are involved in area problems.

➡ **EXAMPLE 11**

Find the area of a rectangle 12mm by 1.5cm   **a** in square centimetres,
                                               b in square millimetres.

a The rectangle is 1.2cm by 1.5cm, so its area is $1.2 \times 1.5 = \underline{1.8 \text{ sq.cm.}}$

b The rectangle is 12mm by 15mm, so its area is $12 \times 15 = \underline{180 \text{ sq.mm.}}$

Note that the 180 is 100 times bigger than 1.8.
To change square centimetres to square millimetres we multiply by 100,
because there are 100 square millimetres in a square centimetre.

Similarly there are 10,000 square centimetres in a square metre,
and 144 square inches in a square foot.

 EXERCISE 8

Find the area of the rectangle in the units stated in brackets:

a 3m by 50cm (sq.m.)   **b** 3m by 50cm (sq.cm.)   **c** 3.2m by 0.2m (sq.m.)

d 3.2m by 0.2m (sq.cm.)   **e** 2ft by 5ft (sq.in.)   **f** 1.2m by 1.2cm (sq.mm.)

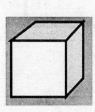

26

VOLUME

The volume of a solid or hollow object is the amount of space inside it.

Volume is measured in cubic units. For example it may be measured in cubic centimetres, cubic metres, cubic feet and so on.

A cubic centimetre is a cube all of whose sides are 1cm in length.

➡️ **EXAMPLE 1**

How many of these   in this?

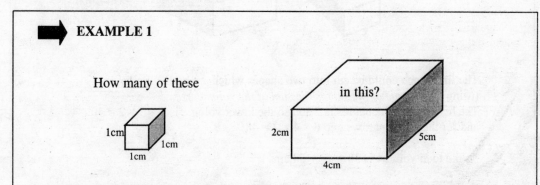

The base of the cuboid on the right here has an area of 20 sq.cm. and so 20 of the cubes on the left will fit onto it.
And since the cuboid is 2cm high there will be 2 layers of 20, making <u>40 cubes</u>.

So the volume of the cuboid is 40 cubic centimetres, written 40 cm^3.

You will see that the answer above is found by multiplying the numbers 4, 5 and 2 together and that this will always give the volume of a cuboid.

> The volume of a cuboid is **Length × Breadth × Height.**
>
> OR   Volume of a cuboid is **Base Area × Height.**

 EXERCISE 1

Find the volume of the figures below (all units are m):

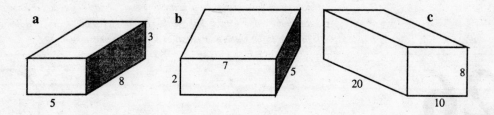

➡ **EXAMPLE 2**

Find the volume of these shapes, all sizes being in cm.

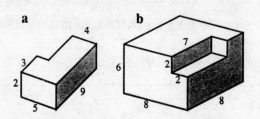

a   The first shape could be cut into two shapes which are both cuboids
(using *the volume of the sum is the sum of the volumes*).
The height of both cuboids is 2 and so the lower volume is $5 \times 3 \times 2 = 30$,
and for the upper part we get $6 \times 4 \times 2 = 48$.

So the total volume is $30 + 48 = \underline{78 \text{ cm}^3}$.

b   Here we could use *By the Non-Completion* and find the volume of the large cuboid
and take the volume of the missing part away.

$8 \times 8 \times 6 - 2 \times 7 \times 2 = 384 - 28 = \underline{356 \text{ cm}^3}$.

? EXERCISE 2

Find the volumes of the figures below (all units are cm):

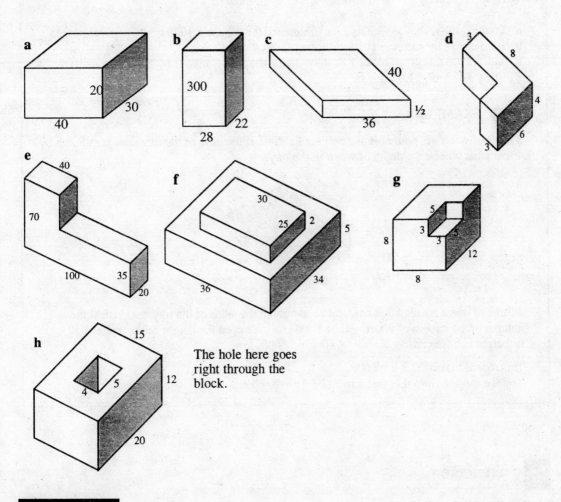

The hole here goes
right through the
block.

CAPACITY

The word capacity has the same meaning as volume but it is used where volumes of liquid are
concerned.

Some units of capacity are the pint, gallon, litre and millilitre.

You should know that   8 pints = 1 gallon,
                            1000 millilitres = 1 litre,
                            1 cm³ = 1 ml. (1 ml means 1 millilitre)

EXAMPLE 3

A carton of fruit juice has dimensions 10cm by 17cm by 6cm.
a   What is its volume in cm³?
What is its capacity   **b** in millilitres,   **c** in litres?

a   Treating the carton as a cuboid its volume is $10 \times 17 \times 6 = 1020$ cm³.
b   1020 cm³ is the same as <u>1020 millilitres</u>.
c   Since 1000 millilitres make 1 litre we divide the above answer by 1000 to convert to
     litres: 1020 ml = <u>1.02 litres</u>.

 EXAMPLE 4

8 litres of water are poured into a rectangular tray. If the base of the tray measures 50cm by
40cm what will be the depth of water in the tray?

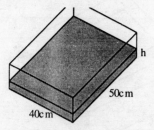

8 litres of water means 8000 cm³ and so assuming the sides of the tray are vertical the
volume of the cuboid of water will be 8000 cm³. We need the height of this cuboid. If
its height is h we can say that $50 \times 40 \times h = 8000$.

This means that $2000 \times h = 8000$.
And we can see from this that h must be 4. So <u>h = 4 cm</u>.

?   **EXERCISE 3**

a Find the capacity of a carton measuring 12cm by 5cm by 16cm   **i** in ml,
                                                                       ii in litres.

b An underground petrol tank has dimensions 6m by 450cm by 200cm.
     Find **i** its volume,
           ii its capacity in ml
           iii its capacity in litres.

c How many litres will a cuboid with dimensions 20cm, 30cm, 40cm contain?

d How many litres will be contained in a cuboid 2m by 2m by 3m?

e 6 litres of water is poured into a rectangular tray with base dimensions 40cm by 50cm. What will be the depth of water in the tray?

f 160 litres of oil is poured into a rectangular container whose base is 4m by 20cm. What will be the depth of the oil?

g Describe a method you could use to find the volume of an irregular shaped container.

*"The astronomer discovers that geometry,
a pure abstraction of the human mind,
is the measure of planetary motion."*

Emerson (1803-82)

AMERICAN PHILOSOPHER AND POET

27

PLANETS

Our solar system consists of 9 planets circling the sun and various other objects.
Some planets are small compared with other much larger planets.
In this chapter you will get an idea of the relative sizes by drawing a circle to represent each planet.

PLANET SIZES

✳ Write a title at the top of a fresh page in your book: <u>Planet Sizes</u>.

✳ In your book put a dot about 5cm from the top of the page and about 3cm from the left side.

✳ With this point as centre draw a circle with radius 1.3cm.
This represents the size of the Earth. Write the name by the circle.

The Earth is actually nearly 13000 kilometers in diameter, or about 8000 miles.

The table below shows the size (in cm) of the other planets on this scale.
1 Using the same centre as for the Earth draw and label a circle for each planet.

For the last 4 planets you will not be able to draw the whole circle so just draw as much as you can.

Mars	Venus	Mercury	Neptune	Uranus	Saturn	Jupiter	
0.7	1.2	0.5	4.8	5.2	12	14.3	(cm)

The scale for this drawing is: 1cm represents 5000km.

On this scale the radius for Pluto is 0.3cm. You can draw this in if you wish but it is not easy to draw such a small circle.

The Moon will also be 0.3cm radius.

The radius of the Sun would be 140cm!

2 Which are the two largest planets?

3 Which planet is about the same size as the Earth?

Since 1cm represents 5000km and 5000km = 500,000,000cm this scale can also be written as a ratio of 1: 500,000,000.

ORBITS OF PLANETS

The path of a planet around the Sun is called its **orbit.**

Mercury is closest to the Sun, then comes Venus, the Earth, Mars, Jupiter, Saturn, Uranus, Neptune and Pluto.

It is not easy to draw a good diagram showing all these orbits drawn to scale.

Let us start with the Earth's orbit and draw the orbit of Venus inside it.

Put a title at the top of your page: Orbits of Earth, Venus and Mercury.
Mark a point near the centre of your page, where the lines intersect.
This point represents the centre of the Sun. Label it S.

4 Draw a circle centred on S and radius 6cm (note this scale is different to the one you used before).
   This is for the orbit of the planet Earth (label this circle "Earth").

You now want to draw a square inside the circle:
Mark 4 points around this circle: 2 points where the horizontal line through S meets the circle, and 2 points where the vertical line through S meets the circle.

By joining these up in order (use a colour) you will get a square.

With your compass point on S draw a circle which just touches the sides of this square. The circle must not go outside the square.

The sides of the square are tangents to the circle.

This circle shows the orbit of the planet Venus (label this circle "Venus").

By extending your drawing you can obtain the orbit of **Mercury**:
We first draw an octagon inside the first circle.

Since you already have a square in this circle you just need 4 more points between the 4 we have already got.

5 Since your page is covered with small squares you can draw a faint <u>diagonal</u> line through S through the corners of these squares to locate 2 points on the big circle.

Then draw another faint diagonal line perpendicular (at right-angles) to the first one to locate the last 2 points.

Label these points 1 to 8 starting at the top and going clockwise round the circle.

Now draw a line, in colour, from point 1 to point 4.

Notice carefully that this line has missed out 2 vertices (the ones numbered 2 and 3).

From 4 draw another line which misses out 2 vertices, this should land at vertex 7.

From point 7 draw the next line (missing 2 vertices) and continue doing this until you reach vertex 1 again.

When you have drawn 8 of these lines you should have an 8-pointed star.
With S as centre draw a circle which just touches the lines of this star.

This is the orbit of the planet Mercury (label this circle "Mercury").

It is worth noting here that although we have drawn the orbits as circles they are not quite circular, and they are not quite centred on the Sun either.

The actual average distance of the Earth from the Sun is about 150 million km.
And the radius of our Earth circle was 6cm.
✸ So what does 1cm represent on our diagram?

Well if 6cm is 150 million km, then 1cm will be 25 million km.
So the scale this time is: 1cm represents 25 million km.

6 Convert 25 million kilometres to centimetres and hence represent this scale as a ratio.

The radius of the other orbits (in cm) are as follows:

Mars	Jupiter	Saturn	Uranus	Neptune	
9	31	57	115	180	(cm)

Pluto is rather peculiar because its orbit sometimes comes inside that of Neptune.

7 Which orbit, including the ones you have drawn, is closest to the orbit of the Earth?

8 How many km is Mars from the Sun?

9 How many times bigger is the orbit of Neptune than the Earth's orbit?

FLEXAGONS

A flexagon, in its final form, is a hexagonal shape which can be folded in such a way that hidden colours can be revealed. There are different sizes of flexagon, having 3, 6, 12 etc. faces and there is another form based on the square.

To make a 3-sided hexagon get some triangular paper and draw out the design below where all the equilateral triangles are of side 4cm. Do not draw the letters.

Front

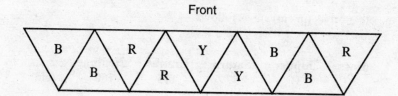

The letters above show how to colour the various triangles on the **front**: B is for Blue, R is for Red and Y is for Yellow. You may use other colours if you wish (or B, R, Y could represent different designs if you do not have colours).

✳ Colour the triangles on the front. Be careful to follow the scheme shown above.

✳ Cut the shape out and fold carefully along every line.
   Turn the shape over and colour in the back according to the scheme shown below. It is important that this is right.

Back

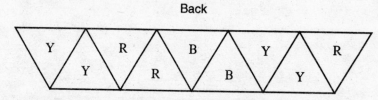

✳ Next we can fold the flexagon up as follows: we are going to hide all the **red** triangles so fold the two adjacent red triangles, which you can see on one side, onto each other.
   Then on the back you will see two more red triangles next to each other. Fold these onto each other as well. You should now have a hexagon shape.
   Hide the red triangles which are left by bringing the back triangle to the front.
✳ Finally put a piece of sellotape on the outside of the hexagon to fix together those two triangles you last moved.

The flexagon is operated by making alternate folds go up and down so that 3 of the folds go up and 3 go down. By squeezing the triangles together the centre can be persuaded to open out, revealing the missing colour.

This process can then be repeated.

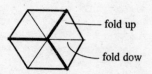

fold up

fold dow

Making flexagons with more hidden faces is similar but not as easy.
Here are directions if you would like to make one with 6 different colours/faces.

Draw out the following on triangular paper using 4cm for each triangle side.

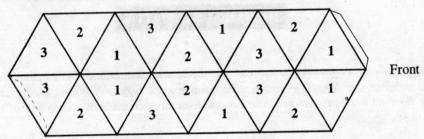

Numbers are used here instead of letters so you will need a colour scheme for 6 colours:
    1 = red, 2 = blue etc, for example.

You will see that there is a small circle near the edge one of the triangles. Make a similar mark as this indicates one of the edges that are sellotaped together at the very end.
* Colour in the front.
* Cut out the shape including the tab.

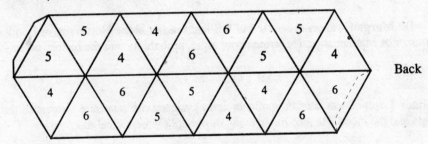

* Fold along the lines, mark the second small circle as shown and colour in the back according to the scheme above. When you turn over for the back turn the left to the right (not the top to the bottom).

The dashed line shows where the tab will fit.
* Cut along the horizontal line and stick the two rows of triangles together to form a continuous line of 18 triangles.

* Now with the back facing you fold the strip into a sort of spiral: fold the 5's on near one end onto each other, then fold the 6's onto each other, then the 4's, the 5's, the 6's, the 4's, the 5's and the 6's.

We are going to hide all sides except sides 1 and 2.
If you next fold the end triangle that has 1 on it so that sides 3 and 4 meet you will be left with a shape just like you had for the first hexagon.

* So fold the pair of 3's together on one side and then fold the pair of 3's on the other side together.
    You now have a hexagon. Arrange all the 1's facing you, then bring the bottom end triangle up and sandwich it between the triangles above it. Sellotape together the edges indicated by the small circles.

HISTORICAL NOTES

Maharishi Mahesh Yogi is the founder of the world wide Transcendental Meditation movement and discoverer of the Absolute Number. He said that the Sutras of Vedic Mathematics are the software for the Cosmic Computer. The Cosmic Computer controls the behaviour of the universe on every scale and level.

Carl Friedrick **Gauss** is regarded as one of the greatest mathematicians of all time. Among many other things he laid the foundations of number theory and rediscovered the lost asteroid Ceres in 1801 using new methods of calculation.

Augustus **De Morgan** was an eccentric but brilliant teacher at the University of London where he was Professor of Mathematics. He wrote about logic, probability and the history of mathematics.

Pierre Simon **Laplace** was a mathematician and physicist with a special interest in the motion of the planets and the moon. He also made many important mathematical discoveries.

Bhāratī Kṛṣṇa **Tīrthajī** rediscovered Vedic Mathematics from ancient Vedic texts between 1911 and 1918. He wrote 16 books but only one is now available. He was a brilliant scholar and was appointed Shankaracharya (religious leader) in 1921. He sought world peace and unity between the East and the West.

Pythagoras founded a society in southern Italy in the late 16th century BC. He is said to have discovered the natural relationships between the notes of the musical scale. An important theorem is also named after him. He believed that the character of each whole number has a cosmic significance.

Plato is one of the most famous of the ancient Greek philosophers. He was a disciple of Socrates and founded an Academy at Athens. He believed that ideas are more real than the physical world and often used mathematics to illustrate his teachings.

Euclid was a Greek mathematician who taught at Alexandria in Egypt. He put together much of the geometrical knowledge established at that time in a textbook in 13 volumes called the *Elements*. This textbook began with Definitions, Postulates and Common Notions and evolved step by step from there deducing many mathematical results.

Bertrand **Russell** was a British philosopher, mathematician and reformer. He tried unsuccessfully to establish the foundations of mathematics in logic.

D'Arcy Thompson brought many new ideas to the study of biology. His book *On Growth and Form* (1917) shows how fundamental mathematical laws and geometrical forms determine the behaviour and evolution of living creatures. He was also a famous conversationalist and lecturer.

Ernst **Kummer** was a German mathematician and scientist. His work in number theory, geometry and physics was extremely original and he was a gifted and humorous teacher.

Pappus of Alexandria was a geometer in ancient Greece. He wrote commentaries on the work of earlier mathematicians (Euclid, Archimedes, Apollonius and others).

Ralph Waldo **Emerson** was an American philosopher and poet. His philosophy was very influential, especially in America.

Jean Le Rond **d'Alembert** was a French mathematician and scientist. He had wide interests and wrote on music and astronomy as well as mathematics.